ECONOMIC OPERATION OF POWER SYSTEMS

LEON K. KIRCHMAYER

Manager, System Generation Analytical Engineering
Electric Utility Engineering Section
General Electric Company
Schenectady, New York

One of a series written by General Electric authors
for the advancement of engineering practice

WILEY EASTERN LIMITED
New Delhi Bangalore Bombay Calcutta

First U.S. Edition, 1958
First Wiley Eastern Reprint, 1979

Authorized reprint of the edition published by
John Wiley & Sons, Inc., New York, London, Sydney and Toronto
Copyright © 1958, John Wiley & Sons, Inc.
All rights reserved. No part of this book may be reproduced
in any form without the written permission of Wiley-Interscience, Inc.

Sales Area: India

This book has been published on the paper supplied through
the Government of India at concessional rate

10 9 8 7 6 5 4 3 2 1

Price: Rs 16.00

ISBN 0 85226 470 4

Published by Ravi Acharya for Wiley Eastern Limited, 4835/24
Ansari Road, Daryaganj, New Delhi 110002 and printed by
Pramodh Kapur at Raj Bandhu Industrial Company, C 61
Maya Puri II, New Delhi 110064. Printed in India.

To
OLGA

PREFACE

This book treats of new analytical and computing techniques which have resulted in significant direct annual savings in the production economy of electric utilities. The emphasis is upon theoretical developments and computer methods which supplement the practical skills of the electric utility engineer.

The methods discussed here have been widely applied by electric utilities in both the United States and Canada. Improvements in fuel economy of approximately $50,000 per year per 1000 mw of installed capacity have been achieved in a number of systems by including the effects of transmission losses by means of the transmission-loss formulas given.

Several powerful tools are described which should greatly enhance the engineer's capability of solving system problems. These are

1. Matrix methods.
2. Analogue and digital computers.

Matrix methods are used to derive and calculate transmission-loss formulas. These matrix methods provide general methods of dealing with systems in a simple manner. They also lead to orderly computational procedures which are readily handled by digital computers. The application of analogue and digital computers to the problems of calculating transmission-loss formulas and generation schedules is also treated.

The application of many of the principles incorporated in this book to automatic economic control of power systems is covered in my book *Economic Control of Interconnected Systems,* now in preparation. This companion work also considers energy accounting, control, and economic theories for economic operation of interconnected systems as well as the subject of economic operation of interconnected steam and hydroelectric systems.

The present volume is based upon a course I have been giving since 1952 to participants in the General Electric Company's Power Systems

Engineering Course. It is intended that the material contained here be suitable for special senior courses or for use as a graduate course in power systems engineering. A knowledge of algebra, calculus, and elementary circuit theory is presumed.

Chapter 1, entitled Introduction, provides a brief historical review of developments leading to improvements in the economic operation of power systems. To the reader who is new to power systems engineering it is suggested that this chapter be read last, since an appreciation of the historical development will not be acquired until the reader has become familiar with the vocabulary of the power systems engineer.

Our research in this area of endeavor was undertaken with the enthusiastic support of S. B. Crary. I wish to thank Mr. Crary sincerely for this encouragement.

Of particular significance to the developments presented here is the fundamental work of G. Kron concerning the theory of transmission-loss formulas. I especially wish to acknowledge the contributions of A. F. Glimn, G. W. Stagg, C. D. Galloway, and R. Habermann, Jr., to the material of this book and to thank Professor J. J. Skiles for his helpful comments in reviewing the manuscript. I also wish to thank my secretary Isabel Godlewski for her excellent work in the preparation of the manuscript.

LEON K. KIRCHMAYER

Schenectady, New York
July 1958

CONTENTS

CHAPTER
1 Introduction — 1
2 Characteristics and Economic Operation of Steam Plants — 8
3 Development of Transmission Loss Formula — 48
4 The Practical Calculation of Loss Formula Coefficients — 116
5 Coordination of Incremental Production Costs and Incremental Transmission Losses for Optimum Economy — 160
6 Practical Calculation, Evaluation, and Application of Economic Scheduling of Generation — 186
7 Transmission Losses as a Function of Voltage Phase Angle — 218
8 Description of Alternative Coordination Methods — 238
9 Evaluation of Energy Differences in the Economic Comparison of Alternative Facilities — 252
Index — 257

1 INTRODUCTION

1.1 SCOPE AND OBJECTIVES

The scope of this book, as indicated by the Table of Contents, includes the theory and practical applications involved in determining the economic operation of a power system. It develops the necessary circuit and mathematical techniques as required in addition to describing the important role that modern computers can play in improving power-system performance. These computing and analytical techniques will be of increasing value to the electric utility engineer, since systems have tended to become progressively complex as they expand. Many of the economic, mathematical, and computing concepts developed in this book are applicable to other fields of endeavor.

1.2 WHY CONSIDER TRANSMISSION LOSSES?

Part of this book concerns methods of calculating transmission losses through means of transmission-loss formulas. In the past much effort has been expended in the analysis of fuel costs and the thermal performance of generating units at equal incremental fuel costs.[1] However, with the development of integrated power systems and the interconnection of operating companies for purposes of economy interchange, it is necessary to consider not only the incremental fuel costs but also the *incremental transmission losses* for optimum economy. The consideration of incremental transmission losses through means of transmission-loss formulas has resulted in fuel savings up to one hundred dollars per year per megawatt of installed peak capacity.

Another important problem in the operation of interconnected systems is the determination of transmission losses for purposes of billing in various interconnection transactions. The revenue to be gained by properly billing for losses involved during interconnection transactions may be a very large sum. For example, the value of the losses incurred by one large operating company in supplying and wheeling power to certain atomic energy loads was of the order of magnitude of one million dollars per year.

2 ECONOMIC OPERATION OF POWER SYSTEMS

Transmission-loss considerations have often proved to be important in the planning of future systems in particular regard to location of plants and the building of transmission lines.

1.3 BRIEF HISTORICAL REVIEW

A transmission-loss formula expressing the total transmission losses in terms of source powers was first presented by E. E. George [2] in 1943. This formula was of the following form:

$$P_L = \text{total transmission losses}$$
$$= B_{11}P_1^2 + B_{22}P_2^2 + B_{33}P_3^2 + \cdots + B_{nn}P_n^2$$
$$+ 2B_{12}P_1P_2 + 2B_{13}P_1P_3 + \cdots$$
$$+ 2B_{23}P_2P_3 + \cdots 2B_{mn}P_mP_n \qquad (1\text{-}1)$$
$$= \sum_m \sum_n P_m B_{mn} P_n$$

P_n, P_m = source loadings

B_{mn} = transmission-loss-formula coefficients

The determination of the B_{mn} coefficients was based on a longhand procedure which required two to three weeks' work by two men for a system of eight to ten generators. Present methods require approximately five per cent or less of this time.

The application of the network analyzer to determine a similar loss formula was developed later by Ward, Eaton, and Hale [3] of Purdue University and published in 1950. However, for the number of generators normally encountered in system studies the arithmetic calculations involved were prohibitively large.

At the 1951 AIEE Summer Convention G. Kron, in conjunction with G. W. Stagg and L. K. Kirchmayer, presented companion papers [4,5] which described an improved method of deriving a total transmission-loss formula requiring considerably less network-analyzer measurements and arithmetic calculations. Reference 4, in addition, evaluated the discrepancies introduced by the assumptions made in obtaining a loss formula.

The application of automatic digital computers to calculate a loss formula was presented by A. F. Glimn, R. Habermann, Jr., L. K. Kirchmayer, and G. W. Stagg in the summer of 1953.[6]

W. R. Brownlee [7] has indicated a method of expressing transmission losses in terms of generator voltages and angles and the X/R ratios of the transmission circuits. Also, loss formulas involving linear terms and a constant term, in addition to the quadratic terms indicated by equation 1-1, have been recently described.[8,9] The form of the loss equation is then

$$P_L = \sum_m \sum_n P_m B_{mn} P_n + \sum_n B_{no} P_n + B_{oo} \qquad (1\text{-}2)$$

This form of loss formula allows more flexibility in the assumptions relating to the manner in which each individual load varies with the total load.

The first major step in the development of a method of coordinating incremental fuel costs and incremental transmission losses was presented in 1949 by E. E. George, H. W. Page, and J. B. Ward [10] in their use of the network analyzer to prepare predicted plant loading schedules for a large power system. At the same time the electrical engineering staff of the American Gas and Electric Service Corporation, also with the aid of the network analyzer, developed a method of modifying the incremental fuel costs of the various plants on an incremental slide rule in order to account for transmission losses. Next, the American Gas and Electric Service Corporation, in cooperation with the General Electric Company, successfully employed transmission-loss formulas and punched-card machines for the preparation of penalty-factor charts to be used in the economic scheduling of generation.[11] The incremental production cost of a given plant multiplied by the penalty factor for that plant gives the incremental cost of power delivered to the system load from that plant. Optimum economy with the effect of transmission losses considered is then obtained when the incremental cost of delivered power is the same from all sources.

In 1952 a paper entitled "Evaluation of Methods of Coordinating Incremental Fuel Costs and Incremental Transmission Losses" [12] presented

1. A mathematical analysis of various methods of coordinating incremental fuel costs and incremental transmission losses.

2. An evaluation of the errors introduced in optimum system operation by assumptions involved in determining a loss formula.

3. An evaluation of the savings to be obtained by coordinating incremental fuel costs and incremental transmission losses.

Progress in the analysis of the economic operation of a combined thermal and hydroelectric power system was reported by the Hydro-Electric Power Commission of Ontario and the General Electric Company in the paper "Short-Range Economic Operation of a Combined Thermal and Hydroelectric Power System," [13] which was presented at the 1953 AIEE Pacific General Meeting.

An iterative method of calculating generation schedules suitable for the use of a high-speed automatic digital computer has been described in the paper entitled "Automatic Digital Computer Applied to Generation Scheduling," by A. F. Glimn, R. Habermann, Jr., L. K. Kirchmayer, and R. W. Thomas.[14] For a given total load, the computer calculates and tabulates incremental cost of received power, total transmission losses, total fuel input, penalty factors, and received load, along with the allocation and summation of generation. Since the program of calculation is general, a single routine is maintained in the computer library which will permit scheduling of any size system in the most efficient manner.

The American Gas and Electric Service Corporation early in 1955 installed an incremental transmission loss computer in their Columbus Production and Coordination Office specifically for the use of the system load dispatcher.[15,16] This computer calculates incremental transmission losses and penalty factors for various system operating conditions. The coordinated operation of this computer and an incremental cost slide rule furnishes a flexible and accurate method of taking into account the various and rapidly changing system conditions in the plant and on the transmission system. Other computer developments include analogue dispatching computers which incorporate both plant incremental cost representation and penalty-factor computation within the computer.[17,18,19]

Rapid progress has been made by the industry in developing economic automation schemes by means of which system frequency, net interchange, and economic allocation of generation for a given area are simultaneously and automatically maintained.[20,21,22,23,24,25,26,27,28] These devices offer important opportunities for savings by

1. Improvement in fuel economy by closer adherence to the optimum schedule than would be possible by manual operation.

2. Possible saving in man-hours by elimination of certain manual procedures.

For operation with the foregoing systems, the net interchange out of the area is set manually and is determined by contracts and bargaining with neighboring areas. Reference 29 describes means of automatically determining and controlling the most economic interchange between areas.

With respect to interconnected systems energy accounting, methods [30,31,32,33,34,35] recently developed permit determination of loss formulas for each separate company and analytical interconnection of the loss formulas of these companies for study of losses in interconnected operation. Loss formulas can be derived that express losses in a given area in terms of the source and interconnection loadings of that area or in terms of source loadings in all areas and the scheduled interchange between areas. These formulas can also be used to express change in losses from a given condition in terms of the changes in source loadings and scheduled flows from a given condition.

1.4 TRENDS IN GROWTH OF POWER SYSTEMS

Because of the following trends in the growth of power systems it has become progressively important to give increasing attention to economic system operation.

1. In many cases economic factors and the availability of primary essentials, such as coal, water, etc., dictate that new generating plants be located at greater distances from the load centers.

2. The installation of larger blocks of power has resulted in the necessity of transmitting power out of a given area until the load in that area is equal to the new block of installed capacity.

3. Power systems are interconnecting for purposes of economy interchange and reduction of reserve capacity.

4. In a number of areas of the country the cost of fuel is rapidly increasing.

References

1. *Economic Loading of Steam Power Plants and Electric Systems*, M. J. Steinberg, T. H. Smith. John Wiley and Sons, New York, 1943.
2. Intrasystem Transmission Losses, E. E. George. *AIEE Trans.*, Vol. 62, March 1943, pp. 153–158.
3. Total and Incremental Losses in Power Transmission Networks, J. B. Ward, J. R. Eaton, H. W. Hale. *AIEE Trans.*, Vol. 69, Part I, 1950, pp. 626–631.
4. Analysis of Total and Incremental Losses in Transmission Systems, L. K. Kirchmayer, G. W. Stagg. *AIEE Trans.*, Vol. 70, Part I, 1951, pp. 1197–1205.
5. Tensorial Analysis of Integrated Transmission Systems—Part I: The Six Basic Reference Frames, G. Kron. *AIEE Trans.*, Vol. 70, Part I, 1951, pp. 1239–1248.
6. Loss Formulas Made Easy, A. F. Glimn, R. Habermann, Jr., L. K. Kirchmayer, G. W. Stagg. *AIEE Trans.*, Vol. 72, Part III, 1953, pp. 730–735.
7. Coordination of Incremental Fuel Costs and Incremental Transmission Losses by Functions of Voltage Phase Angles, W. R. Brownlee. *AIEE Trans.*, Vol. 73, Part III, 1954, pp. 529–541.
8. A General Transmission Loss Equation, E. D. Early, R. E. Watson, G. L. Smith. *AIEE Trans.*, Vol. 74, Part III, 1955, pp. 510–516.
9. A New Method of Determining Constants for the General Transmission Loss Equation, E. D. Early, R. E. Watson. *AIEE Trans.*, Vol. 74, Part III, 1955, pp. 1417–1421.
10. Coordination of Fuel Cost and Transmission Loss by Use of the Network Analyzer to Determine Plant Loading Schedules, E. E. George, H. W. Page, J. B. Ward. *AIEE Trans.*, Vol. 68, Part II, 1949, pp. 1152–1160.

11. Transmission Losses and Economic Loading of Power Systems, L. K. Kirchmayer, G. H. McDaniel. *General Electric Review*, Schenectady, New York, Vol. 54, No. 10, October 1951, pp. 1152–1163.
12. Evaluation of Methods of Coordinating Incremental Fuel Costs and Incremental Transmission Losses, L. K. Kirchmayer, G. W. Stagg. *AIEE Trans.*, Vol. 71, Part III, 1952, pp. 513–521.
13. Short-Range Economic Operation of a Combined Thermal and Hydroelectric Power System, W. G. Chandler, P. L. Dandeno, A. F. Glimn, L. K. Kirchmayer. *AIEE Trans.*, Vol. 72, Part III, 1953, pp. 1057–1065.
14. Automatic Digital Computer Applied to Generation Scheduling, A. F. Glimn, R. Habermann, Jr., L. K. Kirchmayer, R. W. Thomas. *AIEE Trans.*, Vol. 73, Part III-B, 1954, pp. 1267–1275.
15. A Transmission Loss Penalty Factor Computer, C. A. Imburgia, L. K. Kirchmayer, G. W. Stagg. *AIEE Trans.*, Vol. 73, Part III-A, 1954, pp. 567–570.
16. Design and Application of Penalty Factor Computer, C. A. Imburgia, L. K. Kirchmayer, G. W. Stagg, K. R. Geiser. *Proceedings of the American Power Conference*, Vol. XVII, 1955, pp. 687–697.
17. An Incremental Cost of Power-Delivered Computer, E. D. Early, W. E. Phillips, W. T. Shreve. *AIEE Trans.*, Vol. 74, Part III, 1955, pp. 529–534.
18. A Computer for Economic Scheduling and Control of Power Systems, C. D. Morrill, J. A. Blake. *AIEE Trans.*, Vol. 74, Part III, 1955, pp. 1136–1141.
19. Loss Evaluation—Part V: Economic Dispatch Computer Design, R. B. Squires, H. W. Colborn, R. T. Byerly, W. R. Hamilton. *AIEE Trans.*, Vol. 75, Part III, 1956, pp. 719–727.
20. Load Scheduling Goes Automatic, D. H. Cameron, W. S. Burt. *Electrical World*, April 18, 1955, pp. 125–126.
21. A New Type Automatic Dispatching System at Kansas City, D. H. Cameron, E. L. Mueller. *ASME Paper 56-A-215*.
22. Automatic Load-Frequency Control System for Central Station Power, Gustave Ehrenberg. *AIEE Trans.*, Vol. 74, Part III, 1955, pp. 787–795.
23. A New Automatic Dispatching System for Electric Power Systems, K. N. Burnett, D. W. Halfhill, B. R. Shepard. *AIEE Trans.*, Vol. 75, Part III, 1956, pp. 1049–1054.
24. How Modern Tools Improve Power System Operations, L. K. Kirchmayer. *General Electric Review*, Schenectady, New York, September 1956.
25. The "Early Bird" Goes Automatic, E. J. Kompass. *Control Engineering*, December 1956, pp. 77–83.
26. Economic Aspects of General Electric Automatic Dispatching System at Kansas City, D. H. Cameron, E. L. Mueller. *AIEE Conference Paper 57-144* presented at the Winter General Meeting, January 1957.
27. Automatic Economic Dispatching and Load Control, R. H. Travers. *AIEE Trans.*, Vol. 76, Part III, 1957, pp. 291–301.
28. Differential Analyzer Aids Design of Electric Utility Automatic Dispatching System, L. K. Kirchmayer. *AIEE Conference Paper* presented at the Feedback Control Conference, Computers in Control, at Atlantic City, New Jersey, October 1957.
29. Automatic Operation of Interconnected Areas, A. F. Glimn, L. K. Kirchmayer, H. H. Chamberlain. *AIEE Conference Paper* presented at the Summer and Pacific General Meeting, San Francisco, California, June 1956.
30. Analysis of Losses in Interconnected Systems, A. F. Glimn, L. K. Kirchmayer, G. W. Stagg. *AIEE Trans.*, Vol. 71, Part III, 1952, pp. 796–808.

31. Tensorial Analysis of Integrated Transmission Systems—Part III: The Primitive Division, G. Kron. *AIEE Trans.*, Vol. 71, Part III, 1952, pp. 814–822.
32. Tensorial Analysis of Integrated Transmission Systems—Part IV: The Interconnection of Transmission Systems, G. Kron. *AIEE Trans.*, Vol. 72, Part III, 1953, pp. 827–838.
33. Analysis of Losses in Loop-Interconnected Systems, A. F. Glimn, L. K. Kirchmayer, G. W. Stagg. *AIEE Trans.*, Vol. 72, Part III, 1953, pp. 944–953.
34. Power Losses in Interconnected Transmission Networks, H. W. Hale. *AIEE Trans.*, Vol. 71, Part III, 1952, pp. 993–998.
35. Improved Method of Interconnecting Transmission Loss Formulas, A. F. Glimn, L. K. Kirchmayer, J. J. Skiles. *AIEE Paper* presented at Great Lakes District Meeting, East Lansing, Michigan, May 1958.

2 CHARACTERISTICS AND ECONOMIC OPERATION OF STEAM PLANTS

2.1 DEFINITION OF TERMS

To define the terms involved, simplified performance curves of a given turbine-generator boiler unit are drawn in Figures 2.1, 2.2, and 2.3. In Figure 2.1 the fuel input in Btu per hour is plotted as a function of out-

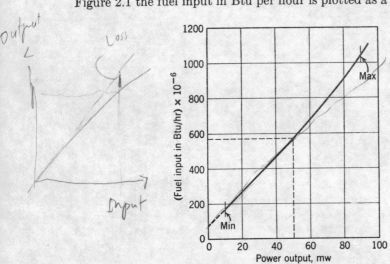

Figure 2.1. Input-output curve.

put in megawatts. The corresponding heat rate, which is obtained by dividing the input by the corresponding output, is given in Figure 2.2. It is to be noted that the units associated with heat rate are Btu per kw-hr. The incremental fuel rate is given by the following definition:

$$\text{incremental fuel rate} = \frac{\Delta \text{ input}}{\Delta \text{ output}} \quad (2\text{-}1)$$

In other words, the incremental fuel rate is equal to a small change in the input divided by the corresponding small change in the output. As the Δ quantities become progressively smaller, it is seen that the

$$\text{incremental fuel rate} = \frac{d \text{ (input)}}{d \text{ (output)}} \tag{2-2}$$

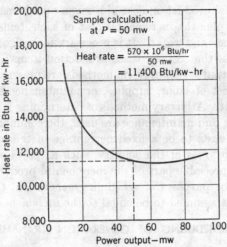

Figure 2.2. Heat rate characteristic.

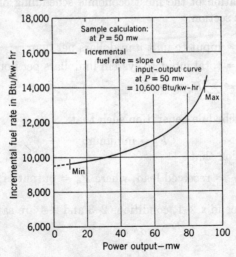

Figure 2.3. Incremental fuel rate curve.

The units associated with the incremental fuel rate are Btu per kw-hr and are the same as the heat-rate units. The incremental fuel rate is converted to incremental fuel cost by multiplying the incremental fuel rate in Btu per kw-hr by the fuel cost in cents per million Btu. The incremental fuel cost is usually expressed in mills per kw-hr or dollars per mw-hr.

2.2 INCREMENTAL PRODUCTION COSTS

The incremental production cost of a given unit is made up of incremental fuel cost plus the incremental cost of such items as labor, supplies, maintenance,[1] and water. It is necessary for a rigorous analysis to be able to express the costs of these production items as a function of instantaneous output. However, no methods are presently available for expressing the cost of labor, supplies, or maintenance accurately as a function of output. Arbitrary methods of determining incremental costs of labor, supplies, and maintenance are used, the commonest of which is to assume these costs to be a fixed percentage of the incremental fuel costs. In certain areas of the country, such as parts of Texas, water costs form an appreciable part of the incremental production costs. In many systems, for purposes of scheduling generation, the incremental production cost is assumed to be equal to the incremental fuel cost.

2.3 OPTIMUM SCHEDULING OF GENERATION (TRANSMISSION LOSSES NEGLECTED)

The determination of the most economic scheduling of generation is discussed in this section.

Let

F_n = input to unit n in dollars per hour

F_t = total input to system in dollars per hour

Then $\quad F_t = \sum_n F_n$

It is desired to schedule generation such that

$$F_t = \text{minimum} \qquad (2\text{-}3)$$

with the restriction that

$$\sum_n P_n = P_R = \text{received load, where } P_n = \text{output of unit } n \qquad (2\text{-}4)$$

As shown in Appendix 2–1, conditions 2–3 and 2–4 are satisfied when

$$\frac{dF_n}{dP_n} = \lambda \qquad (2\text{-}5)$$

where

$\dfrac{dF_n}{dP_n}$ = incremental production cost of unit n in dollars per mw-hr

λ = incremental cost of received power in dollars per mw-hr

The value of λ must be chosen such that $\sum P_n = P_R$.

ECONOMIC OPERATION OF STEAM PLANTS

Stated in terms of words, the minimum input in dollars per hour for a given total load is obtained when all generating units are operated at the same incremental production cost. In equation 2–5 increasing λ results in an increase in total generation; decreasing λ results in a decrease in total generation.

This same result, as in equation 2–5, could be obtained intuitively. Assume that all units are not operating at the same incremental cost. Consequently, some sources would be operated at higher incremental costs than others. It would then be possible to decrease the dollars-per-hour input to the system by increasing the generation at the lower incremental-cost sources and decreasing the generation at the higher incremental-cost sources. In the limiting case it will be seen that all sources should be operated at the same incremental costs.

An alternative method of analysis for a two-source system follows:

$$F_t = F_1 + F_2 \tag{2-6}$$

$$P_1 + P_2 = P_R \quad \text{or} \quad P_2 = P_R - P_1 \tag{2-7}$$

It is desired to obtain the values of P_1 and P_2 that will result in a minimum value of F_t for a given value of P_R.

The value of F_t will be a minimum, according to the methods of calculus, when the first derivative of F_t with respect to P_1 is zero. Thus,

$$\frac{dF_t}{dP_1} = 0 \tag{2-8}$$

Also,
$$\frac{dF_t}{dP_1} = \frac{d(F_1 + F_2)}{dP_1} = \frac{dF_1}{dP_1} + \frac{dF_2}{dP_1} \tag{2-9}$$

From equation 2–7

$$dP_2 = -dP_1$$

since $\quad dP_R = 0$

Hence
$$\frac{dP_1}{dP_2} = -1 \tag{2-10}$$

Combining equations 2–10 and 2–9

$$\frac{dF_t}{dP_1} = \frac{dF_1}{dP_1} - \frac{dF_2}{dP_1}(-1) = \frac{dF_1}{dP_1} - \frac{dF_2}{dP_1}\frac{dP_1}{dP_2}$$

$$= \frac{dF_1}{dP_1} - \frac{dF_2}{dP_2}$$

From equation 2-8, it follows that

$$\frac{dF_t}{dP_1} = \frac{dF_1}{dP_1} - \frac{dF_2}{dP_2} = 0$$

and
$$\frac{dF_1}{dP_1} = \frac{dF_2}{dP_2} \qquad (2\text{-}11)$$

or that the incremental cost of sources 1 and 2 are equal to each other.

Another suggested approach [2] to this problem is to study the effect of arbitrarily swinging generation between these two sources. If the load is assumed constant, then

$$\Delta P_1 + \Delta P_2 = 0 \qquad (2\text{-}12)$$

Assume ΔP_1 is a positive number; that is, that the output of source 1 is increased over its initial amount. Then ΔP_2 will, of course, be equal in magnitude but of opposite sign to ΔP_1.

The input to source 1 will then be increased by an amount ΔF_1; similarly, the input to source 2 will then be decreased by an amount ΔF_2. The change in total fuel input will then be

$$\Delta F_t = \Delta F_1 + \Delta F_2 \qquad (2\text{-}13)$$

Three possible observations may be made concerning ΔF_t:

1. If $\Delta F_t < 0$, the fuel input to the system is decreased by increasing generation on source 1; and, consequently, the initial generation schedule was not the optimum.

2. If $\Delta F_t > 0$, the fuel input to the system is increased by increasing the generation of source 1; and in this case it would not be desirable to increase the output of source 1. In fact it may prove to be desirable to decrease the output of unit 1.

3. If $\Delta F_t = 0$, no improvement is obtained by taking up a small increment of generation on source 1.

Theoretically, any deviation from the optimum loading would result in an increase in fuel input in dollars per hour. However, for practical purposes, the total cost varies slowly with changes from the minimum cost point, and criterion 3 may be used to a precision within the size of the increment ΔP_1. This criterion becomes precisely correct as ΔP_1 approaches zero.

From criterion 3

$$\Delta F_1 + \Delta F_2 = 0 \qquad (2\text{-}14)$$
$$\Delta F_1 = -\Delta F_2$$

ECONOMIC OPERATION OF STEAM PLANTS

Dividing through by ΔP_1,

$$\frac{\Delta F_1}{\Delta P_1} = -\frac{\Delta F_2}{\Delta P_1} \qquad (2\text{-}15)$$

From equation 2-12

$$-\Delta P_2 = \Delta P_1$$

Then

$$\frac{\Delta F_1}{\Delta P_1} = \frac{\Delta F_2}{\Delta P_2}$$

and in the limiting case

$$\frac{dF_1}{dP_1} = \frac{dF_2}{dP_2} \qquad (2\text{-}11)$$

for optimum economy.

2.4 INCREMENTAL FUEL RATE SLIDE RULE

The scheduling of generation for operation at equal incremental production costs may be efficiently accomplished by the use of an incremental-cost slide rule. Figure 2.4 schematically illustrates an incremental-cost slide rule for a two-generator system.

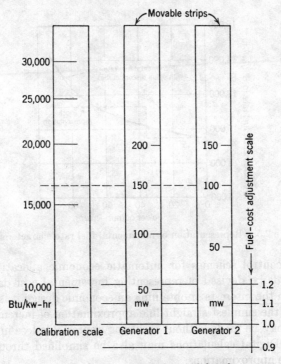

Figure 2.4. Incremental-cost slide rule.

This slide rule consists essentially of a logarithmetic calibration scale, a movable strip for each generator unit, and a fuel-cost adjustment scale. The calibration scale is graduated in Btu per kw-hr to a logarithmic scale. Each movable generator strip is calibrated in megawatts and indicates the relation between the incremental fuel rate and output of a given generator unit. Differences in fuel costs may be accounted for by displacing a given generator strip so as to line up the bottom of the strip to a position on the fuel-cost adjustment scale corresponding to the ratio of fuel costs. For a given incremental cost of received power, corresponding generator outputs can then be read directly from the strips. If the fuel costs for both generators are assumed equal, it is seen from Figure 2.4 that for a total load of 250 mw, 150 should be scheduled from generator 1 and 100 from generator 2.

2.5 APPROXIMATION OF INCREMENTAL PRODUCTION COSTS

Practices vary greatly among companies in the representation of the incremental production costs. Some companies have refined the curve in Figure 2.3 to include discontinuities due to valves, as indicated in Figure 2.5, whereas others use a block representation, also indicated in Figure 2.5.

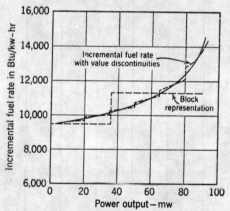

Figure 2.5. Representation of incremental fuel rate characteristics.

Various control schemes for automatic economic allocation require that methods be devised of representing incremental cost data by analogues. In the interests of obtaining an economic design it is necessary to discover the simplest straight-line approximation of incremental cost data that may be made without incurring appreciable loss in operating economy. Digital calculations may also be simplified through use of straight-line approximations.

ECONOMIC OPERATION OF STEAM PLANTS 15

The cost increase resulting from straight-line approximations of incremental cost data are evaluated in this section by consideration of the incremental rate characteristics of Figures 2.6 and 2.7. The cost of fuel is assumed to be the same for both units.

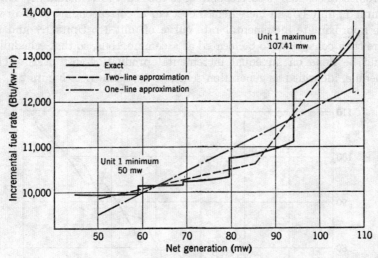

Figure 2.6. Incremental fuel rate for unit 1.

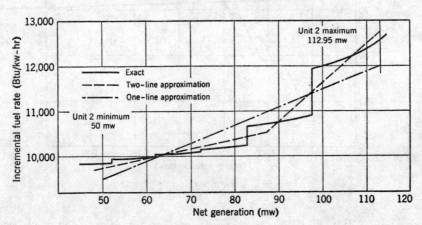

Figure 2.7. Incremental fuel rate for unit 2.

COMPARISON 1

The exact incremental rate curve for unit 1 and the two-line approximation thereof (Figure 2.6) is used here. The approximation was made so as to make the area under each curve the same when integrated from P_1 minimum (50 mw) to P_1 maximum (107.41) and attempting insofar as possible to keep the difference in ordinates to a minimum.

Two units, $1A$ and $1B$, were then postulated. Unit $1A$ is to have an incremental cost curve corresponding to the exact incremental rate curve of unit 1; unit $1B$, an incremental cost curve corresponding to the two-line approximate incremental rate curve of unit 1. Units $1A$ and $1B$ were then considered to be located at a common bus, so that scheduling could be done on an equal incremental production-cost basis. This schedule, identified as generation schedule 1, is shown as Figure 2.8.

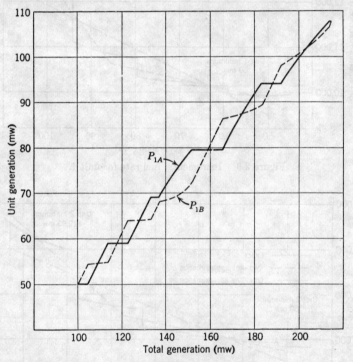

Figure 2.8. Schedule 1.

This schedule is then compared with the schedule that would result if unit $1B$ were to have the same incremental cost curve as unit $1A$ or, in other words, if the incremental cost curves of both units were to be exact. In this case both units would supply one half the total generation, so that no plot of this schedule, which is called generation schedule 2, is required.

ECONOMIC OPERATION OF STEAM PLANTS

It can now be seen that, for a given total generation, unit generation will depend on whether schedule 1 or schedule 2 is followed. This difference in schedules is the result of using an approximation for incremental cost in the case of unit $1B$. Schedule 2 represents the most economic allocation of generation. Therefore, the use of schedule 1 will, in general, specify an allocation of generation resulting in an hourly fuel input in excess of that required if schedule 2 were used. The example which follows illustrates how the resulting hourly increase in fuel input may be calculated.

Assume a total generation of 149 mw. Schedule 1 specified that for this condition unit $1A$ shall supply 78 mw and unit $1B$ shall supply 71 mw (78 + 71 = 149). Schedule 2, however, specifies that the total generation shall be divided equally so that unit $1A$ and unit $1B$ shall each provide 74.5 mw (74.5 + 74.5 = 149). Thus the use of schedule 1 rather than schedule 2 causes the generation at unit $1A$ to be increased from 74.5 to 78 mw, whereas at unit $1B$ generation is decreased from 74.5 to 71 mw. The increase in Btu per hour resulting from the increase of unit $1A$ generation is then (78.0 − 74.5) times (Btu/mw-hr) where (Btu/mw-hr) is taken to be the average incremental Btu per mw-hr in the interval from $P_{1A} = 74.5$ mw to $P_{1A} = 78.0$ mw. Referring to Figure 2.6, (Btu/mw-hr) can be found from the exact incremental rate curve to be equal to $10{,}280 \times 10^3$. Thus, Δ (Btu/hr), due to the increase in unit $1A$ generation, is $10{,}280 \times 10^3 \times 3.5 = +35.980 \times 10^6$. Similarly, the Δ (Btu/hr), due to the decrease in unit $1B$ generation, can be found to be $(71 - 74.5)(10{,}254)(10^3) = -35.889 \times 10^6$. The net increase in Btu per hour caused by the use of schedule 1 rather than schedule 2 is then Δ Btu/hr due to change in unit $1A$ generation + Δ Btu/hr due to change in unit $1B$ generation = $(\Delta \text{ Btu/hr})1_A + (\Delta \text{ Btu/hr})1_B$ = $(35.980)(10^6) - (35.889)(10^6) = 0.091 \times 10^6 = 91{,}000$ Btu/hr. Table 2.1 shows this procedure in tabular form for various values of total generation. Note that at those points where schedules 1 and 2 are identical, there is, of course, no resultant cost increase.

An assumed load-duration curve is given by Figure 2.9. By use of this curve we may plot Btu per hour as a function of the per cent time that the load associated with a particular Btu per hour increase occurs. Thus a plot is made up of Btu per hour vs. per cent time. Such a curve is shown in Figure 2.10. The average of this curve now represents the average increase in Btu per hour caused by the use of an approximation for incremental fuel cost.

Knowing this, we may assign a dollar value to Btu's, say thirty cents per million, and evaluate annual savings. This calculation is shown in Figure 2.10.

ECONOMIC OPERATION OF POWER SYSTEMS

TABLE

P_{1A} Exact Increm. (Schedule 1) mw	P_{1B} Approx. Increm. mw	ΣP $(P_{1A} + P_{1B})$ mw	P_{1A} Exact Increm. (Schedule 2) mw	P_{1B} Exact Increm. mw	ΔP_{1A} $\left(P_{1A} - P_{1A} \atop \text{Sched. 1 Sched. 2} \right)$ mw	$\left(\dfrac{dF}{dP}\right)_{1A}$ Btu/kw-hr
50	50	100	50	50	0	x
50	54.4	104.4	52.2	52.2	−2.2	9,950
54.6	54.6	109.2	54.6	54.6	0	x
59	54.8	113.8	56.9	56.9	+2.1	9,960
59	59	118.0	59	59	0	x
59	63.9	122.9	61.45	61.45	−2.45	10,150
64	64	128	64	64	0	x
69	64.2	133.2	66.6	66.6	+2.4	10,160
69	68	137	68.5	68.5	+0.5	10,160
74	69	143	71.5	71.5	+2.5	10,250
78	71	149	74.5	74.5	+3.5	10,280
79.5	72	151.5	75.75	75.75	+3.75	10,300
79.5	79.5	159	79.5	79.5	0	x
79.5	85.2	164.7	82.35	82.35	−2.85	10,780
79.5	86.4	165.9	82.95	82.95	−3.45	10,785
82.4	86.7	169.1	84.55	84.55	−2.15	10,820
87.4	87.4	174.8	87.4	87.4	0	x
91	88.2	179.2	89.6	89.6	+0.4	10,980
94	89.1	183.1	91.55	91.55	+2.45	11,065
94	94	188	94	94	0	x
94	98	192	96	96	−2.0	12,305
96.8	99	195.8	97.9	97.9	−1.1	12,440
101.5	101.5	203	101.5	101.5	0	x
106	104.8	210.8	105.4	105.4	+0.6	13,100
107.41	106.1	213.51	106.76	106.76	+0.65	13,230
107.41	107.41	214.82	107.41	107.41	0	x

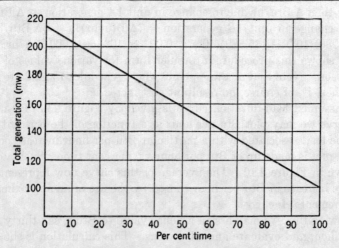

Figure 2.9. Load-duration curve 1.

ECONOMIC OPERATION OF STEAM PLANTS

2.1

$(\Delta P_{1A})\left(\dfrac{dF}{dP}\right)_{1A}$ or $\left(\dfrac{\Delta \text{Btu}}{\text{hr}}\right)_{1A}$ Btu/hr × 10^6	ΔP_{1B} $\left(\begin{array}{cc} P_{1B} & - & P_{1B} \\ \text{Sched. 1} & & \text{Sched. 2} \end{array}\right)$ mw	$\left(\dfrac{dF}{dP}\right)_{1B}$ Btu/kw-hr	$(\Delta P_{1B})\left(\dfrac{dF}{dP}\right)_{1B}$ or $\left(\dfrac{\Delta \text{Btu}}{\text{hr}}\right)_{1B}$ Btu/hr × 10^{-6}	$\left(\dfrac{\Delta \text{Btu}}{\text{hr}}\right)_{1A} + \left(\dfrac{\Delta \text{Btu}}{\text{hr}}\right)_{1B}$ Btu/hr	Time (See load duration curve of Figure 2.9)
0	0	x	0	0	100
−21.890	+2.2	9,950	+21.890	0	96.2
0	0	x	0	0	92
+20.916	−2.1	9,960	−20.916	0	93.5
0	0	x	0	0	84.5
−24.868	+2.45	10,150	+24.868	0	80
0	0	x	0	0	75.6
+24.384	−2.4	10,155	−24.372	12,000	71
+ 5.080	−0.5	10,160	− 5.080	0	67.8
+25.625	−2.5	10,245	−25.613	12,000	62.5
+35.980	−3.5	10,254	−35.889	91,000	57.3
+38.625	−3.75	10,264	−38.490	135,000	55
0	0	x	0	0	48.6
−30.723	+2.85	10,827	+30.857	134,000	43.7
−37.208	+3.45	10,845	+37.415	207,000	42.7
−23.263	+2.15	10,865	+23.360	97,000	39.7
0	0	x	0	0	35
+ 4.392	−0.4	10,937	− 4.375	17,000	31
+27.109	−2.45	10,983	−26.908	201,000	27.5
0	0	x	0	0	23.2
−24.610	+2.0	12,420	+24.840	230,000	19.8
−13.684	+1.1	12,510	+13.761	77,000	16.5
0	0	x	0	0	10.2
+ 7.860	−0.6	13,045	− 7.827	33,000	3.5
+ 8.600	−0.65	13,180	− 8.567	33,000	1.0
0	0	x	0	0	0

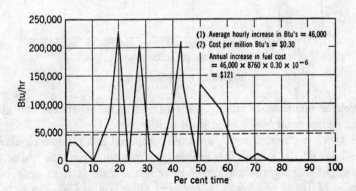

Figure 2.10. Evaluation of increased annual cost of schedule 1 over schedule 2.

COMPARISON 2

This comparison differs from comparison 1 only in that a single straight line is used to approximate the incremental fuel rate of unit 1 rather than the two-straight-line approximation used previously.

Again two units, $1A$ and $1B$, were postulated, with the incremental fuel rate of unit $1A$ being taken as the exact curve of Figure 2.6 and that of unit $1B$ being considered as the one-straight-line approximation also shown by Figure 2.6. The resulting allocation of generation is given as Figure 2.11, schedule 3. This schedule is then compared with schedule

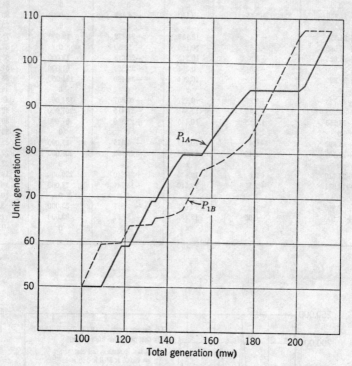

Figure 2.11. Schedule 3.

2, which, as mentioned earlier, is based on using an exact representation for the incremental fuel cost of both units.

The two schedules, 2 and 3, are then compared in a manner similar to Table 2.1, and the annual increase in cost resulting from the use of the one-straight-line approximation of incremental rate is then evaluated. The plot of Btu's per hour vs. per cent time is given as Figure 2.12, which shows the evaluation of the annual cost increase involved.

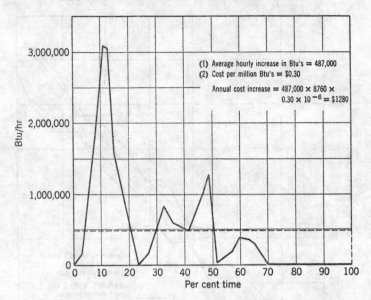

Figure 2.12. Evaluation of increased annual cost of schedule 3 over schedule 2.

COMPARISON 3

In both comparison 1 and comparison 2 the two units being scheduled were chosen to be identical, although in some instances approximate characteristics were used rather than the exact ones. Now, for comparison 3 we shall consider the scheduling of two different units, first using the exact incremental fuel rate for each unit and then a two-line approximation for each unit.

Figure 2.6 gives the exact and approximate incremental-rate data for unit 1, and Figure 2.7 gives the same information for unit 2. Figure 2.13 shows two schedules, 4 and 5. Schedule 4 shows the allocation of generation between units 1 and 2 when scheduling is done on the basis of equal incremental cost with the exact incremental-cost data used for both units. Schedule 5 shows how generation would be divided between units 1 and 2 when equal incremental-cost scheduling is followed but with the incremental-rate data for each plant being represented by a two-straight-line approximation.

The difference in cost resulting from the two schedules is then computed in a manner similar to that outlined for comparisons 1 and 2. Since unit 2 has a maximum rating greater than that of unit 1, a new load-duration curve is assumed. This curve, load-duration curve 2, is shown as Figure 2.14.

22 ECONOMIC OPERATION OF POWER SYSTEMS

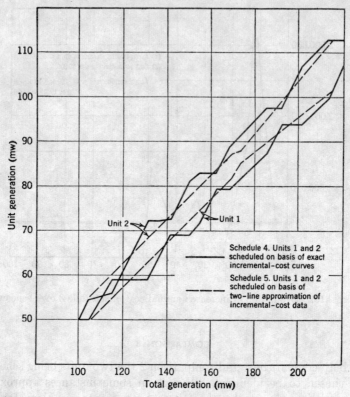

Figure 2.13. Schedules 4 and 5.

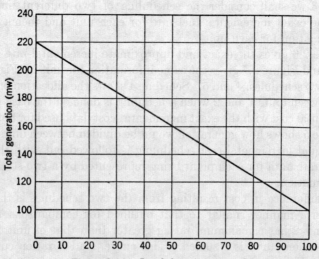

Figure 2.14. Load-duration curve 2.

Figure 2.15 shows the determination of annual cost increase due to the scheduling of units 1 and 2 by means of two-line approximations of the incremental production costs of each unit.

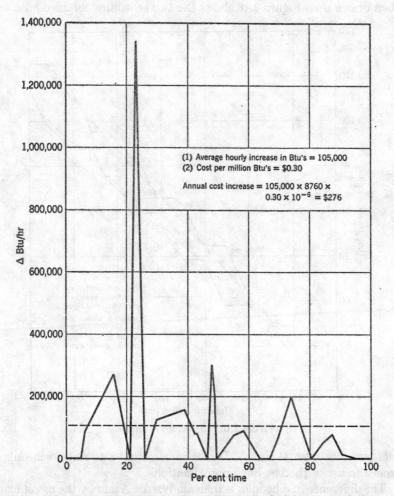

Figure 2.15. Evaluation of increased annual cost of schedule 5 over schedule 4.

COMPARISON 4

This calculation is similar to comparison 3, except that when units 1 and 2 are scheduled on an approximate basis a one-line approximation of incremental fuel rate is assumed rather than the two-line approximation used before.

Exact incremental-rate data and the one-line approximation of this data is given by Figures 2.6 and 2.7 for units 1 and 2, respectively. Schedules based on equal incremental-production-cost scheduling are then drawn up. Figure 2.16 shows the two schedules obtained. Sched-

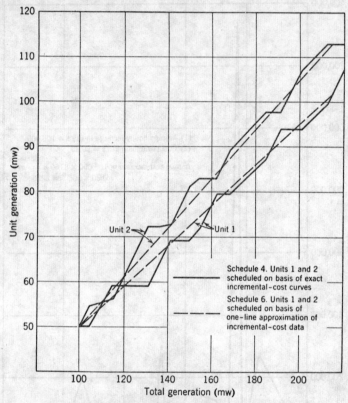

Figure 2.16. Schedules 4 and 6.

ule 4 results from the use of exact incremental-rate data; schedule 6 from the use of the one-line approximations.

The difference in schedules is than analyzed. Again by the use of load-duration curve 2 (Figure 2.14), an annual cost increase resulting from the use of approximations for incremental fuel cost is evaluated. This evaluation is shown by Figure 2.17.

From the results on page 26, it appears that a one-line approximation for incremental-fuel-cost curves of the type of Figure 2.6 is not sufficiently accurate for purposes of economic scheduling. For this particular problem, the use of a two-line approximation seems to be sufficiently accurate for practical purposes.

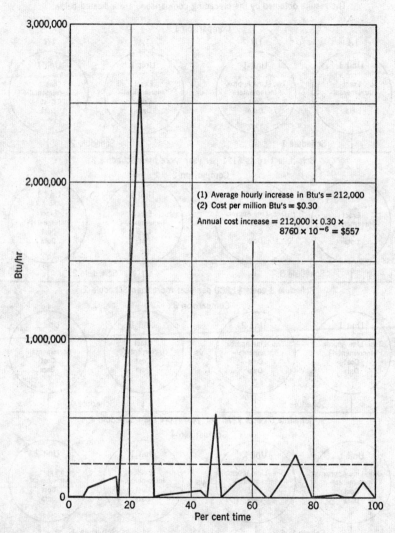

Figure 2.17. Evaluation of increased annual cost of schedule 6 over schedule 4.

The results obtained by the preceding comparisons are indicated below:

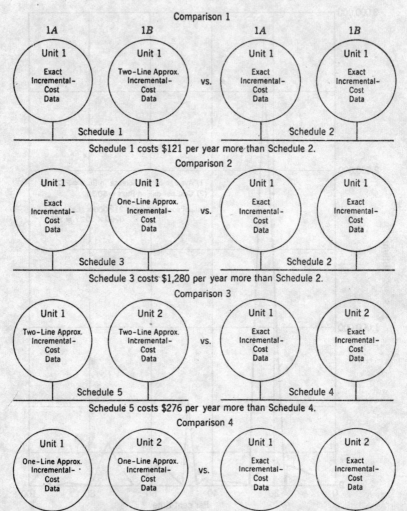

2.6 OTHER METHODS OF SCHEDULING TURBINE-GENERATORS

Although the criterion of equal incremental production costs will result in the optimum economic scheduling of generation, the following methods of scheduling are sometimes still found in use.[3]

1. *Base Loading to Capacity.* The turbine-generators are successively loaded to capacity in order of their efficiencies.
2. *Base Loading to Most Efficient Load.* The turbine-generator units are successively loaded, in ascending order of their heat rates, to their most efficient loads. When all units are operating at their most efficient loads they are loaded to capacity in the same order.
3. *Proportional to Capacity.* The loads on the units are scheduled in proportion to their rated capacity.

2.7 DETERMINATION OF CAPACITY TO BE OPERATED

The discussion thus far has considered the optimum allocation of generation for a given connected capacity. A problem which is not answered by inspection of incremental-cost data is the determination of the capacity to be operated for a given total load. Determination of this capacity is based upon such considerations as

1. Economic evaluation.
2. Reserve requirements.
3. Stability limitations.
4. Voltage limitations.
5. Ability to pick up load quickly.

Very frequently, and in particular in widespread systems, conditions 2 to 5 overrule condition 1.

The determination of the most economic combination of capacity to be operated at a given time is accomplished by inspection of the total fuel input to the system for various assumed combinations of capacity. Of course, for any assumed capacity in operation the economic allocation of generation is given by equal incremental-cost loading.

In general, in a given station the units are placed in service in ascending order of their heat rates assuming the cost per Btu to be the same. To determine the most economic combination of units for a given station load it is necessary to plot total station heat-rate curves of successive combinations and to note the combination providing the lowest heat rate for a given station load.

Another problem of importance is to determine the economic advisability of taking units off the line for relatively short periods of time, such as between the morning and evening peaks. This determination is based upon calculating the total fuel input in dollars to the system during this period of time with the units in question both on and off the line. This calculation should include cost of restoring the units under consideration back in service and losses involved in banking the boilers.

2.8 BRIEF DESCRIPTION OF A TYPICAL MANUAL PROCEDURE IN LOAD DISPATCHER'S OFFICE

Load forecasts for the expected hourly loadings of a given day are usually prepared on the preceding day. The capacity available is dependent upon the maintenance schedule for the system as well as any forced outages that may have occurred. The choice of units to be operated from this available capacity is discussed in Section 2.7.

Based upon the forecasted hourly loads and the capacity that is to be operated, the load dispatcher then determines hourly generation schedules for the following day's operation by the use of the incremental slide rule. The plant operators are informed of their respective plant schedules and manually follow that particular schedule. In case of variations from predicted conditions the load dispatcher is called upon to revise his precalculated generation schedule.

Recent developments in computers and automatic dispatching systems have removed the need for many of these manual steps. Several dispatching computers are described in Chapter 6. Various automatic methods of obtaining economic allocation of generation, together with the maintenance of frequency and net interchange, are discussed in *Economic Control of Interconnected Systems* by L. K. Kirchmayer (in preparation).

2.9 EFFECTS OF ERRORS IN ECONOMIC DISPATCHING OF POWER SYSTEMS [4]

An analysis of the effects of errors in the economic dispatching of systems is important in understanding and choosing the accuracy requirements of the components of a dispatching system.

Deviations from the most economic schedule are obtained if

1. The representation of the incremental-production-cost curve is in error.
2. The servomechanism loop which matches the desired generation with the actual generation is inaccurate. In the case of manual operation the station operator represents the servomechanism loop.

The two types of error indicated may occur in an automatic dispatching system as well as in the manual dispatching of a power system.

ECONOMIC OPERATION OF STEAM PLANTS

SYSTEM STUDIED

The system considered consists of two 100-mw units, as shown in Figure 2.18. Incremental cost data is given by

Figure 2.18. Schematic representation of system studied.

$$\frac{dF_1}{dP_1} = \text{incremental production cost of unit 1 in dollars per mw-hr}$$
$$= 2 + 0.012 P_1 \qquad (2\text{--}16)$$

$$\frac{dF_2}{dP_2} = \text{incremental production cost of unit 2 in dollars per mw-hr}$$
$$= 1.5 + 0.015 P_2 \qquad (2\text{--}17)$$

Assume that
1. Both units are connected and operating.
2. Minimum load of each unit is 10 mw.
3. Maximum capability of each unit is 100 mw.
4. The total load varies from 50 to 200 mw, as shown in Figure 2.19.

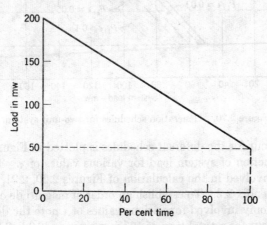

Figure 2.19. Load-duration curve.

ERROR IN REPRESENTATION OF INCREMENTAL-COST DATA

Assume that dF_1/dP_1 is incorrectly represented as being high by a factor $\epsilon[(dF_1)/(dP_1)]$ and that dF_2/dP_2 is represented as being $\epsilon[(dF_2)/(dP_2)]$ lower than the correct value. Consequently, schedules that are different from the most economic schedule are obtained when ϵ is other than zero.

Figure 2.20 shows the correct schedule and modified schedules for units 1 and 2 for various values of ϵ. The increase in production cost in

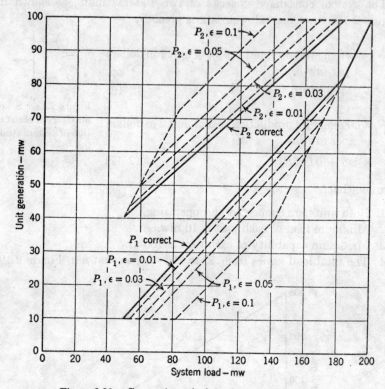

Figure 2.20. Generation schedules for two-unit system.

dollars per hour for the different schedules is plotted in Figures 2.21 and 2.22 as a function of system load for various values of ϵ. The method of analysis involved in the calculation of Figures 2.20, 2.21, and 2.22 is given in Appendix 2.2. To establish an order of magnitude of the deviations in economy involved for various values of ϵ note the deviations in dollars per hour for a total load of 125 mw given in Table 2.2:

TABLE 2.2

ϵ	Loss in Economy in Dollars per Hour
0.00	0.00
0.01	0.05
0.03	0.46
0.05	1.28
0.1	5.18

ECONOMIC OPERATION OF STEAM PLANTS

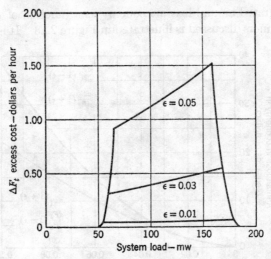

Figure 2.21. Excess costs incurred in deviation from optimum schedules.

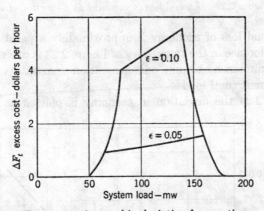

Figure 2.22. Excess costs incurred in deviation from optimum schedules.

If the multipliers $(1 + \epsilon)$ and $(1 - \epsilon)$ are interchanged, then the loss in economy becomes as shown in Table 2.3.

TABLE 2.3

ϵ	Loss in Economy in Dollars per Hour
0	0
0.01	0.05
0.03	0.45
0.05	1.25
0.1	4.94

The integrated annual loss in economy as a function of ϵ assigned in the two manners discussed is illustrated in Figure 2.23. It is to be noted

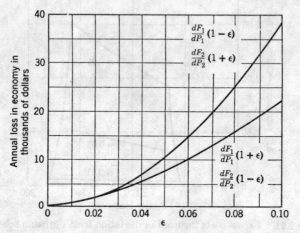

Figure 2.23. Annual loss in economy as a function of ϵ.

that the annual loss of economy is approximately a parabolic function of ϵ. The difference in the two curves of Figure 2.23 is due mainly to the fact that minimum-maximum unit-generation constraints are encountered at different total loads.

In Figure 2.24 the deviation in economy is plotted as the ratio of

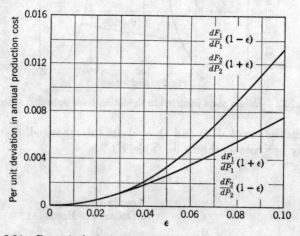

Figure 2.24. Per unit deviation in annual production cost as function of ϵ.

annual loss in economy to the total fuel consumption when operated most economically. The fuel consumption was obtained by integrating the proper incremental-cost curves and adding the assumed no-load production-cost intercepts of twenty dollars per hour and forty dollars per hour, respectively, for units 1 and 2. As noted in Figure 2.24, the per unit deviation in fuel economy is much smaller than the corresponding deviation assumed in the incremental-production-cost data. To assist in obtaining an understanding of this relationship Figure 2.25 plots

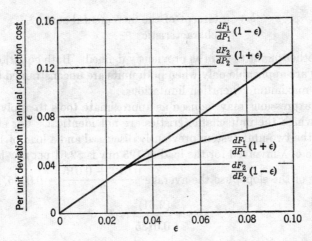

Figure 2.25. Ratio of per unit deviation in annual production cost to ϵ as function of ϵ.

the ratio of the per unit deviation in fuel economy to the deviation in incremental-cost data as a function of the deviation in incremental-cost data.

Appendix 2.3 presents an analytical determination of the loss in operating economy resulting from an incremental-cost displacement multiplier error $(1 + \epsilon)$ for one of two identical units on the same bus and $(1 - \epsilon)$ for the other unit. The hourly loss in economy is given by the expression

$$\Delta F_t = \frac{(\lambda \epsilon)^2}{a} \qquad (2\text{--}18)$$

where ΔF_t = dollars-per-hour loss of operating economy

λ = incremental-cost level for two identical units on the same bus as the load

a = slope of incremental-cost characteristic of each unit

ϵ = deviation in incremental-cost representation

It is seen from equation 2–18 that the deviation of fuel economy varies as the square of the error ϵ.

The loss in operating economy may also be expressed in terms in the total generation as

$$\Delta F_t = \left(\frac{a}{2} P_R + b\right)^2 \frac{\epsilon^2}{a} \tag{2-19}$$

where $\quad b =$ intercept of the incremental-cost characteristic

and the other quantities are as previously defined. Both equations 2–18 and 2–19 are applicable only when both units are unconstrained by minimum or maximum generation limitations.

These expressions may be used as approximate tools to apply to systems in which the unit characteristics are not identical. For example, consider the two-unit system previously discussed and a load of 125 mw. The value of λ involved for the load of 125 mw is \$2.61 per mw-hr. For the value of the slope, use the average $\dfrac{0.012 + 0.015}{2} = 0.0135$.

Then
$$\Delta F_t = \frac{(2.61)^2}{0.0135} (\epsilon)^2$$

$$= 505(\epsilon)^2$$

From the above equation we obtain the following results:

TABLE 2.4

ϵ	Loss in Economy in Dollars per Hour
0	0
0.01	0.05
0.03	0.45
0.05	1.26
0.1	5.05

By comparison of Table 2.4 with Tables 2.2 and 2.3 it will be seen that close agreement is obtained between the results of the approximate formula, as given in Table 2.4, and the results presented in Tables 2.2 and 2.3.

EFFECT OF ERROR IN HOLDING GENERATION AT DESIRED VALUE

Although the incremental-production-cost data may be represented accurately, an error in the unit loading may occur because of errors in the automatic or manual execution of the schedule. For purposes of this discussion assume that the error is a fixed amount of the maximum unit load but that the unit minimums and maximums will still be held at 10 and 100 mw, respectively. A per unit error of 0.1 in loading by this definition corresponds to 0.1 × 100, or a 10-mw shift in loading; i.e., unit 1 loading would then be low by 10 mw and unit 2 loading, high by 10 mw. The loss in economy as a function of various per unit errors is plotted in Figure 2.26 for the two-unit system described. Figure 2.27

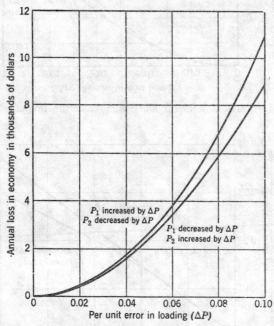

Figure 2.26. Annual loss in economy as a function of error in loading.

presents the same results in terms of the ratio of the annual loss in economy to the production cost for optimum loading as a function of the loading error. Figure 2.28 gives the ratio of the per unit loss in economy to the per unit loading error as a function of the deviation in loading.

Appendix 2.3 presents the derivation of the following expression for the loss of economy resulting from errors in maintaining generation at desired values for the special case of two identical units:

$$\Delta F_t = \Delta P^2 a \tag{2-20}$$

where ΔP = displacement in loading on one of two identical units

a = slope of incremental-cost curve of either unit

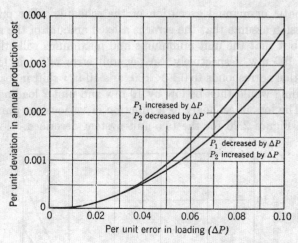

Figure 2.27. Deviation in annual production cost as a function of error in loading.

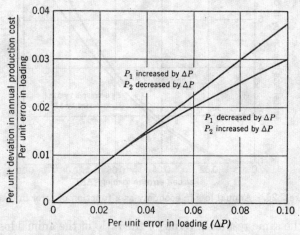

Figure 2.28. Ratio of per unit deviation in annual production cost to error in loading as function of error in loading.

The loss in economy varies as the square of the error in loading. Applying this expression to the small system of two dissimilar units previously discussed, we have

$$\Delta F_t = \Delta P^2 (0.0135)$$

where $\qquad a = 0.0135 =$ average slope

Thus, for a simultaneous deviation of $+10$ mw of one unit and -10 mw

ECONOMIC OPERATION OF STEAM PLANTS

on the other

$$\Delta F_t = (10)^2(0.0135)$$

$$= \$1.35 \text{ per hour}$$

CONCLUSIONS

1. From the study of the effect of simultaneously displacing the incremental cost of one source by $(1 + \epsilon)$ and the incremental cost of another source by $(1 - \epsilon)$ the following conclusions may be drawn for the two-source system investigated: (a) The loss in hourly economy varies as the square of the per unit error ϵ in the representation of the incremental production cost. (b) For a given value of ϵ the loss of hourly economy varies directly as the square of the incremental-cost level and inversely as the average slope of the incremental-cost characteristics of the sources in question.

2. From the study presented of the effect of displacing the output of one source by $+\Delta P$ mw and the output of another source by $-\Delta P$ mw from the desired economic value it is noted that the loss of hourly economy varies as the square of the ΔP mw deviation from optimum schedule.

2.10 SYSTEMS WITH TRANSMISSION LOSSES

In the general case all sources of generation are not located at the same bus but are connected by means of a transmission network to the various loads. Some plants will be more favorably located with respect to the load than others. Also, if the criterion of equal incremental production costs is applied, there will, in general, be transmission of power from low-cost to high-cost areas. It will be necessary, of course, for optimum economic operation to recognize that transmission losses occur in this operation and to modify the incremental production costs of all plants to take these line losses into account.

A simple system [5] that is used as an example is given in Figure 2.29.

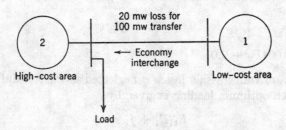

Figure 2.29. Schematic representation of system.

This system is a simple representation of the American Gas and Electric System (1950) and illustrates the relatively high-cost generation in the Indiana Division as compared to the low-cost generation in the Ohio Division available for transfer west. These two areas are interconnected by a 132-kw transmission system and are separated by a distance of approximately 250 miles. The maximum practical transfer over the transmission system corresponds to approximately 170 mw.

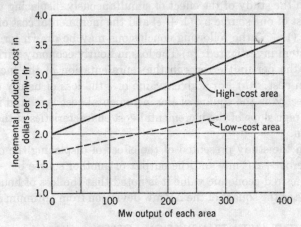

Figure 2.30. Incremental fuel cost data.

The approximate incremental production costs for plants 1 and 2 are shown in Figure 2.30 and may be expressed by

$$\frac{dF_1}{dP_1} = \text{incremental production cost of plant 1 in dollars per mw-hr}$$

$$= F_{11}P_1 + f_1$$

$$= 0.002P_1 + 1.7 \tag{2-21}$$

$$\frac{dF_2}{dP_2} = \text{incremental production cost of plant 2 in dollars per mw-hr}$$

$$= F_{22}P_2 + f_2$$

$$= 0.004P_2 + 2.0 \tag{2-22}$$

If the effect of transmission losses is neglected in the scheduling of generation, then optimum loading is given by

$$F_{11}P_1 + f_1 = \lambda \tag{2-23}$$

$$F_{22}P_2 + f_2 = \lambda$$

ECONOMIC OPERATION OF STEAM PLANTS

Various values of total generation are obtained by varying the value of λ. It will be noted that if the two areas are operated at the same incremental production costs 150 mw will be scheduled from the low-cost area before generation is scheduled from the high-cost area. The division of generation as a function of total generation is given in Figure 2.31. The

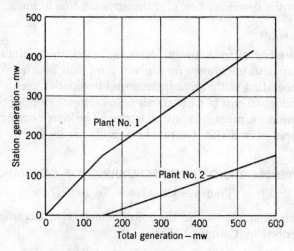

Figure 2.31. Generation schedule.

fuel input as a function of received load with generation scheduled by equal incremental production costs is given in Figure 2.32. However, if

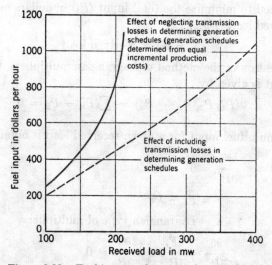

Figure 2.32. Fuel input as function of received load.

the effect of incremental transmission losses had been included in the scheduling of generation, the fuel input would be reduced, as shown in Figure 2.32.

In order to be able to coordinate incremental production costs and incremental transmission losses properly, we shall next study the theory involved in the determination of a transmission-loss formula.

2.11 SUMMARY

With the effect of transmission losses neglected, the minimum dollars-per-hour input to the system for a given total load is obtained when all units are operating at the same incremental production cost. The incremental production cost of a given unit is equal to the slope of the dollars-per-hour input vs. megawatt output curve. The incremental production cost is expressed in dollars per mw-hr or mills per kw-hr.

APPENDIX 2.1. OPTIMUM SCHEDULING OF GENERATION
(Transmission Losses Neglected)

This derivation follows directly from the method of Lagrangian multipliers described by Courant.[6] Let

$$F_t = \text{total input to system in dollars per hour}$$
$$= \sum_n F_n$$

where F_n = input to plant n in dollars per hour

It is desired to minimize the total input (F_t) in dollars per hour for a given received load (P_R). Let

$$P_R = \text{given received load}$$

By application of the method of Lagrangian multipliers, the equation of constraint is given by

$$\Psi(P_1, P_2, P_3 \cdots P_n) = \sum_n P_n - P_R = 0 \qquad (2\text{-}24)$$

Then minimum fuel input for a given received load is obtained when

$$\frac{\partial \mathcal{F}}{\partial P_n} = 0 \qquad (2\text{-}25)$$

where
$$\mathcal{F} = F_t - \lambda \Psi \qquad (2\text{-}26)$$

λ = Lagrangian type of multiplier

$$\frac{\partial \mathcal{F}}{\partial P_n} = \frac{\partial F_t}{\partial P_n} - \lambda \frac{\partial \Psi}{\partial P_n} = 0 \qquad (2\text{-}27)$$

ECONOMIC OPERATION OF STEAM PLANTS 41

Then
$$\frac{\partial F_t}{\partial P_n} - \lambda \frac{\partial}{\partial P_n}\left[\sum_n P_n - P_R\right] = 0$$

$$\frac{\partial F_t}{\partial P_n} - \lambda[1 - 0] = 0$$

$$\frac{\partial F_t}{\partial P_n} = \lambda \qquad (2\text{-}28)$$

But
$$\frac{\partial F_t}{\partial P_n} = \frac{\partial\left(\sum_n F_n\right)}{\partial P_n} = \frac{\partial F_n}{\partial P_n} = \frac{dF_n}{dP_n} \qquad (2\text{-}29)$$

Then equation 2–28 becomes
$$\frac{dF_n}{dP_n} = \lambda \qquad (2\text{-}30)$$

Equation 2–30 is identical to equation 2–5 previously presented.

APPENDIX 2.2. METHOD OF CALCULATION OF LOSS OF ECONOMY

1. Calculation of Schedules

Assume two units whose incremental costs are represented by the following equations:

$$\frac{dF_1}{dP_1} = \text{incremental cost of unit 1 in dollars per mw-hr}$$
$$= F_{11}P_1 + f_1$$

where P_1 = load on unit 1

$$\frac{dF_2}{dP_2} = \text{incremental cost of unit 2 in dollars per mw-hr}$$
$$= F_{22}P_2 + f_2$$

where P_2 = load on unit 2

The most economic schedule for the case of no transmission losses occurs when the incremental plant costs are equal. For this case the most economic schedule is given by

$$\frac{dF_1}{dP_1} = F_{11}P_1 + f_1 = \lambda \qquad (2\text{-}31)$$

$$\frac{dF_2}{dP_2} = F_{22}P_2 + f_2 = \lambda \qquad (2\text{-}32)$$

where λ = incremental cost of received power

From equations 2-31 and 2-32 we obtain

$$P_2 = \frac{F_{11}P_1 + f_1 - f_2}{F_{22}} \qquad (2\text{-}33)$$

Defining P_R as the total load on both units,

$$P_2 = P_R - P_1 \qquad (2\text{-}34)$$

Equating 2-33 and 2-34 and solving for P_1, we obtain

$$P_1 = \frac{F_{22}P_R - f_1 + f_2}{F_{11} + F_{22}} \qquad (2\text{-}35)$$

Values for P_2 and P_1 for various values of P_R may be obtained from equations 2-34 and 2-35, respectively.

Assume that dF_1/dP_1 is multiplied by $(1 + \epsilon)$ and that dF_2/dP_2 is multiplied by $(1 - \epsilon)$. For these modified incremental costs the expression for P_1 then becomes

$$P_1 = \frac{F_{22}P_R - f_1 + f_2 - \epsilon(f_1 + f_2 + F_{22}P_R)}{F_{11} + F_{22} + (F_{11} - F_{22})\epsilon} \qquad (2\text{-}36)$$

The value for P_2 is, of course, given by

$$P_2 = P_R - P_1 \qquad (2\text{-}34)$$

2. Calculation of ΔF_t for Various Values of P_R

Let Figure 2.33 represent the exact incremental cost data. For a given

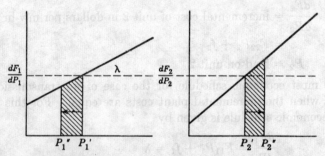

Figure 2.33. Representation of incremental cost data.

incremental cost λ and the corresponding value of P_R let the value of P_1 and P_2 obtained with the exact incremental data be denoted by P_1' and P_2'.

ECONOMIC OPERATION OF STEAM PLANTS

If the modified incremental-cost data is used, different values of P_1 and P_2 will be obtained for the same value of P_R. Denote these values of P_1 and P_2 by P_1'' and P_2''.

In going from P_2' to P_2'' the fuel input to unit 2 is increased by

$$\Delta F_2 = \left[\frac{\left(\dfrac{dF_2}{dP_2}\right)'' + \left(\dfrac{dF_2}{dP_2}\right)'}{2}\right] \Delta P_2 \qquad (2\text{--}37)$$

where $\quad \Delta P_2 =$ change in generation on unit $2 = P_2'' - P_2'$

$$\left(\frac{dF_2}{dP_2}\right)'' = \frac{dF_2}{dP_2} \quad \text{for } P_2 = P_2''$$

$$\left(\frac{dF_2}{dP_2}\right)' = \frac{dF_2}{dP_2} \quad \text{for } P_2 = P_2'$$

The fuel input to unit 1, however, is decreased by

$$\Delta F_1 = \left[\frac{\left(\dfrac{dF_1}{dP_1}\right)'' + \left(\dfrac{dF_1}{dP_1}\right)'}{2}\right] \Delta P_1 \qquad (2\text{--}38)$$

Since the load dropped by one unit must be picked up by the other, the change in P_1 must be equal and opposite to the change in P_2. Thus $\Delta P = -\Delta P_1 = \Delta P_2$.

Then the change in total fuel input is given by

$$\Delta F_t = \Delta F_1 + \Delta F_2$$

$$= \left\{\left[\left(\frac{dF_2}{dP_2}\right)'' + \left(\frac{dF_2}{dP_2}\right)'\right] - \left[\left(\frac{dF_1}{dP_1}\right)'' + \left(\frac{dF_1}{dP_1}\right)'\right]\right\} \frac{\Delta P}{2}$$

but $\left(\dfrac{dF_2}{dP_2}\right)' = \left(\dfrac{dF_1}{dP_1}\right)'$

Hence

$$\Delta F_t = \left[\left(\frac{dF_2}{dP_2}\right)'' - \left(\frac{dF_1}{dP_1}\right)''\right] \frac{\Delta P}{2} \qquad (2\text{--}39)$$

By the use of equation 2–39, ΔF_t may be quickly and accurately calculated.

APPENDIX 2.3. DERIVATION OF EXPRESSION FOR HOURLY LOSS IN OPERATING ECONOMY RESULTING FROM ERROR IN INCREMENTAL-COST REPRESENTATION

It is desired to calculate the loss of operating economy involved when the incremental production cost associated with two identical units is modified.

Let
$$\frac{dF_1}{dP_1} = F_{11}P_1 + f_1$$

and
$$\frac{dF_2}{dP_2} = F_{22}P_2 + f_2$$

where
$$F_{11} = F_{22} = a \tag{2-40}$$

and
$$f_1 = f_2 = b \tag{2-41}$$

Assume that dF_1/dP_1 is multiplied by $(1 + \epsilon)$ and that dF_2/dP_2 is multiplied by $(1 - \epsilon)$. The new schedule obtained is given by solution of

$$\frac{dF_1}{dP_1}(1 + \epsilon) = \frac{dF_2}{dP_2}(1 - \epsilon) \tag{2-42}$$

From equation 2-36 of Appendix 2.2 and equations 2-40 and 2-41 we obtain

$$P_1 = \frac{aP_R - \epsilon(2b + aP_R)}{2a} = \frac{P_R}{2}(1 - \epsilon) - \frac{\epsilon b}{a} \tag{2-43}$$

The shift in loading for each of the two identical units may be computed from the expression

$$\Delta P = \frac{P_R}{2} - P_1 \tag{2-44}$$

$$= \frac{P_R}{2} - \left[\frac{P_R}{2}(1 - \epsilon) - \frac{\epsilon b}{a}\right]$$

$$= \left(\frac{P_R}{2} + \frac{b}{a}\right)\epsilon = \left(\frac{aP_R + 2b}{2a}\right)\epsilon = \left(\frac{\frac{a}{2}P_R + b}{a}\right)\epsilon \tag{2-45}$$

ECONOMIC OPERATION OF STEAM PLANTS

The hourly loss in fuel economy is given by equation 2–39 of Appendix 2.2 as

$$\Delta F_t = \left[\left(\frac{dF_2}{dP_2}\right)'' - \left(\frac{dF_1}{dP_1}\right)''\right]\frac{\Delta P}{2} \qquad (2\text{--}46)$$

but

$$\left(\frac{dF_2}{dP_2}\right)'' = a\left(\frac{P_R}{2} + \Delta P\right) + b$$

$$\left(\frac{dF_1}{dP_1}\right)'' = a\left(\frac{P_R}{2} - \Delta P\right) + b$$

Then

$$\left(\frac{dF_2}{dP_2}\right)'' - \left(\frac{dF_1}{dP_1}\right)'' = 2a\,\Delta P$$

$$\Delta F_t = 2a\,\Delta P\,\frac{\Delta P}{2} \qquad (2\text{--}20)$$

$$= a\,\Delta P^2$$

Substituting equation 2–45 into equation 2–20,

$$\Delta F_t = a\left[\frac{\dfrac{a}{2}P_R + b}{a}\right]^2 \epsilon^2$$

$$= \left[\frac{a}{2}P_R + b\right]^2 \frac{\epsilon^2}{a} \qquad (2\text{--}19)$$

For the case of two identical units, with no transmission losses, we may define the incremental cost of received power as

$$aP_1 + b = \lambda \qquad (2\text{--}47)$$

$$aP_2 + b = \lambda \qquad (2\text{--}48)$$

Adding equations 2–47 and 2–48,

$$a(P_1 + P_2) + 2b = 2\lambda$$

$$\lambda = \frac{a}{2}P_R + b \qquad (2\text{--}49)$$

Substituting equation 2–49 into equation 2–19, we obtain

$$\Delta F_t = \lambda^2 \frac{\epsilon^2}{a} \qquad (2\text{--}18)$$

References

1. Incremental Maintenance Costs of Steam-Electric Generating Stations, M. J. Steinberg, *AIEE Trans.*, Paper 57–1060 presented at Fall Meeting, October 1957.
2. Economy Loading Simplified, J. B. Ward. *AIEE Trans.*, Vol. 72, Part III, 1953, pp. 1306–1311.
3. *Economy Loading of Power Plants and Electric Systems*, M. J. Steinberg, T. H. Smith. John Wiley and Sons, New York, 1943, pp. 89–97.
4. Accuracy Considerations in Economic Dispatching of Power Systems, A. F. Glimn, L. K. Kirchmayer, G. W. Stagg, V. R. Peterson. *AIEE Trans.*, Vol. 75, Part III, 1956, pp. 1125–1137.
5. Evaluation of Methods of Coordinating Incremental Fuel Costs and Incremental Transmission Losses, L. K. Kirchmayer, G. W. Stagg. *AIEE Trans.*, Vol. 71, Part III, 1952, pp. 513–520.
6. *Differential and Integral Calculus*, R. Courant. Interscience Publishers, New York, Vol. 11, 1936, pp. 188–211.

Problems

Problem 2.1

Assume that the fuel inputs in Btu per hour for units 1 and 2 are given by

$$F_1 = (8P_1 + 0.024P_1^2 + 80)10^6$$

$$F_2 = (6P_2 + 0.04P_2^2 + 120)10^6$$

where F_n = fuel input to unit n in Btu per hour

P_n = unit output in megawatts

a. Plot the input-output characteristic for each unit expressing input in Btu per hour and output in megawatts. Assume that the minimum loading of each unit is 10 mw and that the maximum loading is 100 mw.

b. Calculate the heat rate in Btu per kw-hr and plot against output in megawatts.

c. Assume that the cost of fuel is twenty-five cents per million Btu. Calculate the incremental production cost in dollars per mw-hr of each unit and plot against output in megawatts.

Problem 2.2

a. Assume the system load to vary from 50 to 200 mw. For the data of Problem 2.1 plot the outputs of units 1 and 2 as a function of total system load when scheduling generation by equal incremental production costs. Assume that both units are operating.

b. The following procedure will obtain the optimum loading in another manner:
It is desired to determine the optimum generation schedule for a load of 100 mw. Plot the total fuel input for a load of 100 mw as generation is shifted from one unit to the other as indicated in Figure 2.34.

The schedule of generation for minimum fuel input in dollars per hour should check the schedule obtained by equal incremental production costs.

c. Assume that the total load starts at 50 mw and increases to a value of 200 mw. Determine which unit would be placed in service first and also the system load at which the remaining unit should be placed in service.

ECONOMIC OPERATION OF STEAM PLANTS

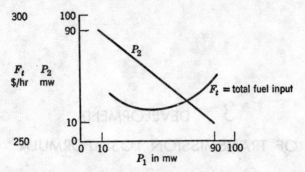

Figure 2.34. Fuel input as function of P_1.

Problem 2.3

The system to be studied consists of the two units described in problems 2.1 and 2.2. Assume a daily load cycle as sketched in Figure 2.35. Also, assume that a cost of thirty dollars is incurred in taking either unit off the line and returning to service after twelve hours.

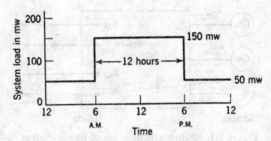

Figure 2.35. Daily load cycle.

Consider the twenty-four-hour period from 6:00 A.M. one morning to 6:00 A.M. the next morning.

a. Would it be more economical to keep both units in service for this twenty-four-hour period or to remove one of the units from service for the twelve-hour period from 6:00 P.M. one evening to 6:00 A.M. the next morning?

b. What is the economic schedule for the period of time from 6:00 A.M. to 6:00 P.M. (load = 150 mw)?

c. What is the economic schedule for the period of time from 6:00 P.M. to 6:00 A.M. (load = 50 mw)?

3 DEVELOPMENT
OF TRANSMISSION LOSS FORMULA

3.1 STATEMENT OF PROBLEM

The basic problem involved is the determination of an expression for the transmission losses in terms of the source loadings. It is desired to proceed from a circuit in which the various sources are connected by an arbitrary transmission network to the individual loads, as indicated in Figure 3.1, to an equivalent circuit, as indicated in Figure 3.2. Figure

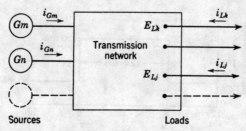

Figure 3.1. Schematic diagram of power system.

3.1 corresponds to the system representation used in network-analyzer studies. The transmission losses [1,2] obtained in both Figure 3.1 and Figure 3.2 are to be identical and may be expressed in the following manner:

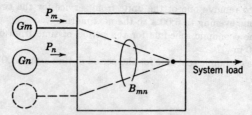

Figure 3.2. Equivalent circuit with impressed generator powers.

For Figure 3.1

$$P_L = \sum_k i_k^2 R_k$$

DEVELOPMENT OF TRANSMISSION LOSS FORMULA

where
$\quad i_k$ = scalar line current in line k
$\quad R_k$ = resistance of line k

For Figure 3.2
$$P_L = \sum_m \sum_n P_m B_{mn} P_n$$

where
$\quad P_m, P_n$ = source loadings
$\quad B_{mn}$ = constants to be determined

In considering the measurements and methods used it is necessary to consider the accuracy limitations of the computer to be used. For example, a number which is obtained by subtracting two approximately equal voltages measured on the network analyzer would be of questionable accuracy.

3.2 INTRODUCTION TO NOTATION

In order to introduce the notation and fundamental concepts involved we shall first consider a simple circuit with two sources, as shown in Figure 3.3. The operation of the circuit may be defined by the following

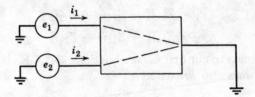

Figure 3.3. Two-source system.

set of equations:

$$e_1 = Z_{11}i_1 + Z_{12}i_2$$
$$e_2 = Z_{21}i_1 + Z_{22}i_2$$

(3-1)

The preceding set of equations may be written as a matrix, as indicated below:

$$\begin{array}{|c|c|} \hline e_1 \\ \hline e_2 \\ \hline \end{array} = \begin{array}{|c|c|} \hline Z_{11} & Z_{12} \\ \hline Z_{21} & Z_{22} \\ \hline \end{array}$$

(3-2)

or

$$\begin{array}{c|c|} & 1 \\ \hline 1 & e_1 \\ \hline 2 & e_2 \\ \end{array} = \begin{array}{c|cc|} & 1 & 2 \\ \hline 1 & Z_{11} & Z_{12} \\ \hline 2 & Z_{21} & Z_{22} \\ \end{array} \begin{array}{c|c|} & 1 \\ \hline 1 & i_1 \\ \hline 2 & i_2 \\ \end{array} \qquad (3\text{-}3)$$

or
$$e_m = Z_{mn}i_n \quad \text{where } m = 1, 2 \qquad (3\text{-}4)$$
$$n = 1, 2$$

It is to be noted that whenever a repeated index appears a summation on that index is indicated. Thus, for the product $Z_{mn}i_n$ a summation on n is indicated. In equation 3-4 let m first equal 1 and then 2. We then have

$$e_1 = Z_{1n}i_n$$
$$e_2 = Z_{2n}i_n \qquad (3\text{-}5)$$

Letting n assume the values of 1 and 2 in each equation, we obtain

$$e_1 = Z_{1n}i_n = Z_{11}i_1 + Z_{12}i_2$$
$$e_2 = Z_{2n}i_n = Z_{21}i_1 + Z_{22}i_2$$

which corresponds to our first set of equations, 3-1. Equations 3-3 and 3-4 are sometimes also written as

$$e = Zi$$

where e, Z, and i are understood to be matrices.

Consider the two-source system in Figure 3.4. We shall consider the

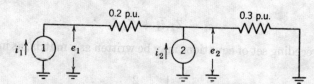

Figure 3.4. Schematic diagram of illustrative two-source system.

manner in which to obtain the values of Z_{mn}. It will be noted from equation 3-3 that if $i_1 = 1$ p.u. and $i_2 = 0$ p.u.

$$\begin{array}{c|c|} & 1 \\ \hline 1 & e_1 \\ \hline 2 & e_2 \\ \end{array} = \begin{array}{c|c|} & 1 \\ \hline 1 & Z_{11} \\ \hline 2 & Z_{21} \\ \end{array}$$

DEVELOPMENT OF TRANSMISSION LOSS FORMULA

Thus, with $i_1 = 1$ p.u. and $i_2 = 0$, the voltages correspond to the self and mutual impedances with respect to source 1.

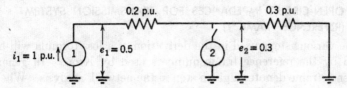

Figure 3.5. Measurement of Z_{m1}.

From Figure 3.5 it will be noted that

$$\begin{array}{|c|} \hline 1 \\ \hline e_1 \\ \hline e_2 \\ \hline \end{array} \begin{array}{c} 1 \\ \end{array} = \begin{array}{|c|} \hline 1 \\ \hline 0.5 \\ \hline 0.3 \\ \hline \end{array} = \begin{array}{|c|} \hline 1 \\ \hline Z_{11} \\ \hline Z_{21} \\ \hline \end{array}$$

Similarly, with $i_1 = 0$ and $i_2 = 1$ p.u., we have as may be noted in Figure 3.6 and in equation 3-3

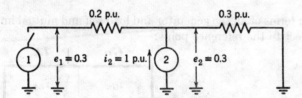

Figure 3.6. Measurement of Z_{m2}.

$$\begin{array}{|c|} \hline 2 \\ \hline e_1 \\ \hline e_2 \\ \hline \end{array} = \begin{array}{|c|} \hline 2 \\ \hline 0.3 \\ \hline 0.3 \\ \hline \end{array} = \begin{array}{|c|} \hline 2 \\ \hline Z_{12} \\ \hline Z_{22} \\ \hline \end{array}$$

Thus

$$\begin{array}{|c|c|} \hline & 1 & 2 \\ \hline 1 & Z_{11} & Z_{12} \\ \hline 2 & Z_{21} & Z_{22} \\ \hline \end{array} = \begin{array}{|c|c|} \hline & 1 & 2 \\ \hline 1 & 0.5 & 0.3 \\ \hline 2 & 0.3 & 0.3 \\ \hline \end{array}$$

It is to be noted that $Z_{12} = Z_{21}$. In general, $Z_{mn} = Z_{nm}$ for static circuits.

3.3 OPEN-CIRCUIT IMPEDANCES FOR TRANSMISSION SYSTEM (REFERENCE FRAME 1)

The various steps used in the derivation of a loss formula will be denoted by the reference frame numbers used by Kron.[1] In general, a reference frame denotes a given step in the network analysis. When the variables are changed from one set to another this change may be thought of as a change in reference frame.

If in Figure 3.1 any given point in the transmission network is chosen as a reference point (Figure 3.7), the following set of equations may be

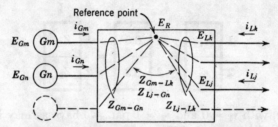

Figure 3.7. Self and mutual impedances for transmission network (reference frame 1).

written in terms of all the generator and load self and mutual impedances with respect to the reference point:

$$
\begin{array}{c|c}
Gn & Lk \\
\hline
i_{Gn} & i_{Lk}
\end{array}
$$

$$
\begin{array}{c|c}
Gm & E_{Gm} - E_R \\
\hline
Lj & E_{Lj} - E_R
\end{array}
=
\begin{array}{c|cc}
 & Gn & Lk \\
\hline
Gm & Z_{Gm-Gn} & Z_{Gm-Lk} \\
\hline
Lj & Z_{Lj-Gn} & Z_{Lj-Lk}
\end{array}
\quad (3\text{-}6)
$$

where
m, n = number of sources
j, k = number of loads

The indices G and L are identification indices referring to generator and load, respectively. Since these equations refer to reference frame 1, they may be written

$$E_1 = Z_{11} I_1 \quad (3\text{-}7)$$

DEVELOPMENT OF TRANSMISSION LOSS FORMULA

The various currents and voltages are defined in Figure 3.7. The equivalent load current i_{Lk} at bus k is defined as the sum of the line-charging, synchronous condenser and load current at that bus.

The impedances designated by Z_{Gm-Gn} represent the self and mutual impedances between the generators. The term Z_{Lj-Lk} represents the self and mutual impedances between the loads; and the terms Z_{Gm-Lk} and Z_{Lj-Gn} represent the mutual impedances between the generators and the loads.

The impedances given in equations 3-6 and 3-7 may be measured as described in the following steps:

1. Remove from ground all line-charging capacitors, synchronous condensers, loads, and generators.
2. Ground the reference point. Generally one of the major plants is chosen as reference point. In many actual systems a prudently chosen reference point results in many zero terms in the impedance matrix.
3. Impress a known current at a given generator n, as in Figure 3.8, and measure all load and generator voltages.

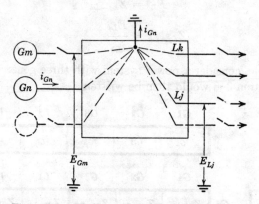

Figure 3.8. Measurement of Z_{Gm-Gn} and Z_{Lj-Gn}.

Then
$$E_{Gm} = Z_{Gm-Gn} i_{Gn}$$

and
$$Z_{Gm-Gn} = \frac{E_{Gm}}{i_{Gn}} \tag{3-8}$$

$$E_{Lj} = Z_{Lj-Gn} i_{Gn}$$

$$Z_{Lj-Gn} = \frac{E_{Lj}}{i_{Gn}} \tag{3-9}$$

4. As shown in Figure 3.9, impress a known current at load Lk and measure all load and generator voltages.

54   ECONOMIC OPERATION OF POWER SYSTEMS

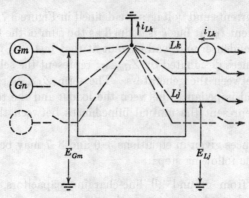

Figure 3.9. Measurement of Z_{Gm-Lk} and Z_{Lj-Lk}.

Then
$$E_{Gm} = Z_{Gm-Lk} i_{Lk}$$

and
$$Z_{Gm-Lk} = \frac{E_{Gm}}{i_{Lk}} \tag{3-10}$$

Also,
$$E_{Lj} = Z_{Lj-Lk} i_{Lk} \tag{3-11}$$

and
$$Z_{Lj-Lk} = \frac{E_{Lj}}{i_{Lk}}$$

For example, consider a power system with three generators and two loads. Equation 3–6 would then be written

	G1	G2	G3	L1	L2
	i_{G1}	i_{G2}	i_{G3}	i_{L1}	i_{L2}

			G1	G2	G3	L1	L2	
G1	$E_{G1} - E_R$	G1	Z_{G1-G1}	Z_{G1-G2}	Z_{G1-G3}	Z_{G1-L1}	Z_{G1-L2}	
G2	$E_{G2} - E_R$	G2	Z_{G2-G1}	Z_{G2-G2}	Z_{G2-G3}	Z_{G2-L1}	Z_{G2-L2}	
G3	$E_{G3} - E_R$ =	G3	Z_{G3-G1}	Z_{G3-G2}	Z_{G3-G3}	Z_{G3-L1}	Z_{G3-L2}	(3-12)
L1	$E_{L1} - E_R$	L1	Z_{L1-G1}	Z_{L1-G2}	Z_{L1-G3}	Z_{L1-L1}	Z_{L1-L2}	
L2	$E_{L2} - E_R$	L2	Z_{L2-G1}	Z_{L2-G2}	Z_{L2-G3}	Z_{L2-L1}	Z_{L2-L2}	

The one-line diagram is given in Figure 3.10. Generator bus $G3$ is chosen as the reference point. If $i_{G1} = 1$ p.u. is impressed, as in Figure 3.11, it will be noted that

DEVELOPMENT OF TRANSMISSION LOSS FORMULA

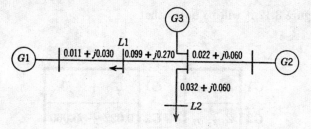

Figure 3.10. Impedance diagram for simple three-source system.

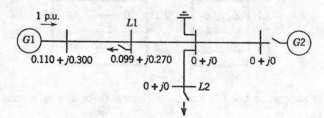

Figure 3.11. Generator $G1$ energized.

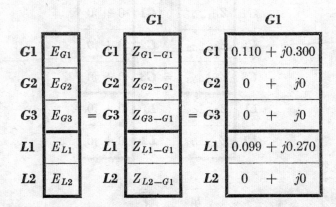

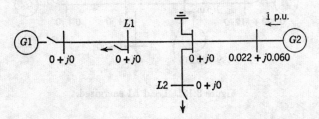

Figure 3.12. Generator $G2$ energized.

From Figure 3.12 it will be noted that

$$
\begin{array}{c|c}
 & G2 \\
\hline
G1 & Z_{G1-G2} \\
G2 & Z_{G2-G2} \\
G3 & Z_{G3-G2} \\
L1 & Z_{L1-G2} \\
L2 & Z_{L2-G2}
\end{array}
=
\begin{array}{c|c}
 & G2 \\
\hline
G1 & 0 + j0 \\
G2 & 0.022 + j0.060 \\
G3 & 0 + j0 \\
L1 & 0 + j0 \\
L2 & 0 + j0
\end{array}
$$

Since the generator bus $G3$ has been chosen to be the reference bus, it follows that

$$
\begin{array}{c|c}
 & G3 \\
\hline
G1 & Z_{G1-G3} \\
G2 & Z_{G2-G3} \\
G3 & Z_{G3-G3} \\
L1 & Z_{L1-G3} \\
L2 & Z_{L2-G3}
\end{array}
=
\begin{array}{c|c}
 & G3 \\
\hline
G1 & 0 + j0 \\
G2 & 0 + j0 \\
G3 & 0 + j0 \\
L1 & 0 + j0 \\
L2 & 0 + j0
\end{array}
$$

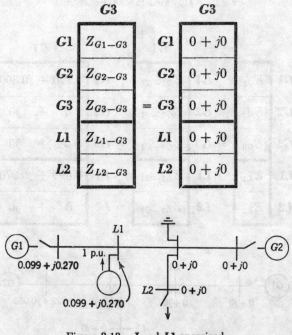

Figure 3.13. Load $L1$ energized.

With $i_{L1} = 1$ p.u. current, it will be noted from Figure 3.13 that

DEVELOPMENT OF TRANSMISSION LOSS FORMULA

	L1			L1
G1	Z_{G1-L1}		G1	$0.099 + j0.270$
G2	Z_{G2-L1}		G2	$0 + j0$
G3	Z_{G3-L1}	=	G3	$0 + j0$
L1	Z_{L1-L1}		L1	$0.099 + j0.270$
L2	Z_{L2-L1}		L2	$0 + j0$

Similarly, the results obtained with $i_{L2} = 1$ p.u. current are

	L2			L2
G1	Z_{G1-L2}		G1	$0 + j0$
G2	Z_{G2-L2}		G2	$0 + j0$
G3	Z_{G3-L2}	=	G3	$0 + j0$
L1	Z_{L1-L2}		L1	$0 + j0$
L2	Z_{L2-L2}		L2	$0.032 + j0.060$

The resultant impedance matrix is

	G1	G2	G3	L1	L2
G1	$0.110 + j0.300$	$0 + j0$	$0 + j0$	$0.099 + j0.270$	$0 + j0$
G2	$0 + j0$	$0.022 + j0.060$	$0 + j0$	$0 + j0$	$0 + j0$
G3	$0 + j0$	$0 + j0$	$0 + j0$	$0 + j0$	$0 + j0$
L1	$0.099 + j0.270$	$0 + j0$	$0 + j0$	$0.099 + j0.270$	$0 + j0$
L2	$0 + j0$	$0 + j0$	$0 + j0$	$0 + j0$	$0.032 + j0.060$

(3–13)

This matrix of self and mutual impedances completely defines the performance of the transmission system of Figure 3.10.

3.4 MULTIPLICATION OF MATRICES [3,4]

Before discussing the concept of transformation, we shall first briefly discuss the multiplication of matrices. As will be recalled from the discussion of equations 3-1, 3-2, and 3-3, the multiplication of a rectangular matrix by a single column matrix is given by

$$
\begin{array}{c}
\rightarrow \\
1\phantom{A_{11}}2 \\
\begin{array}{c}1\\2\end{array}\!\!\begin{bmatrix}A_{11} & A_{12}\\ A_{21} & A_{22}\end{bmatrix}
\end{array}
\times
\begin{array}{c}
\downarrow 1 \\
2
\end{array}\!\!\begin{bmatrix}B_1\\ B_2\end{bmatrix}
=
\begin{array}{c}1\\2\end{array}\!\!\begin{bmatrix}A_{11}B_1 + A_{12}B_2\\ A_{21}B_1 + A_{22}B_2\end{bmatrix}
\qquad (3\text{-}14)
$$

$$A \quad \times \quad B \quad = \quad C$$

If the B matrix is also rectangular, we have

$$A \times B = C$$

$$
\begin{array}{c}1\\2\end{array}\!\!\begin{bmatrix}A_{11} & A_{12}\\ A_{21} & A_{22}\end{bmatrix}
\times
\begin{array}{c}1\\2\end{array}\!\!\begin{bmatrix}B_{11} & B_{12}\\ B_{21} & B_{22}\end{bmatrix}
=
\begin{array}{c}1\\2\end{array}\!\!\begin{bmatrix}A_{11}B_{11}+A_{12}B_{21} & A_{11}B_{12}+A_{12}B_{22}\\ A_{21}B_{11}+A_{22}B_{21} & A_{21}B_{12}+A_{22}B_{22}\end{bmatrix}
$$

$$
=
\begin{array}{c}1\\2\end{array}\!\!\begin{bmatrix}C_{11} & C_{12}\\ C_{21} & C_{22}\end{bmatrix}
\qquad (3\text{-}15)
$$

Let C_{mn} denote the element in row m and column n of matrix C. It will be noted that element C_{mn} from the C matrix is obtained by multiplying the elements in the m row of the A matrix by elements in the n column of the B matrix. Thus, the element C_{12}, as indicated below, is obtained by multiplying the elements in the first row of the A matrix by the elements in the second column of the B matrix.

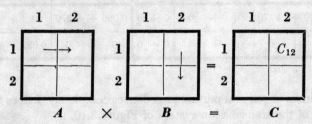

DEVELOPMENT OF TRANSMISSION LOSS FORMULA

From inspection of the arrows in these matrices, the arrow rule of multiplication may be stated in these words:

The terms in the products entering into the determination of C_{mn} may be obtained by drawing a horizontal arrow pointing to the right in the m row of matrix A and a vertical arrow pointing down in the n column of matrix B. In general, if A has m rows and p columns and B has p rows and n columns, then C will have m rows and n columns. This relationship is indicated below for $m = 5$, $p = 3$, and $n = 4$.

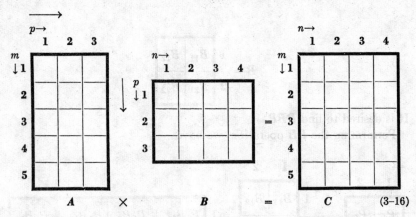

$$A \times B = C \qquad (3\text{-}16)$$

The operation of multiplication of matrices may also be performed in terms of index notation. Corresponding to the matrix multiplication indicated by equation 3-15, we have

$$A_{mp}B_{pn} = C_{mn}$$

The element C_{12} is given by

$$C_{12} = A_{1p}B_{p2}$$
$$= A_{11}B_{12} + A_{12}B_{22}$$

This result is the same as that given by equation 3-15.

If we perform the operation BA, we obtain the following result:

$$\begin{array}{c} \\ 1 \\ 2 \end{array} \begin{array}{|c|c|} \hline B_{11} & B_{12} \\ \hline B_{21} & B_{22} \\ \hline \end{array} \begin{array}{|c|c|} \hline A_{11} & A_{12} \\ \hline A_{21} & A_{22} \\ \hline \end{array} = \begin{array}{|c|c|} \hline B_{11}A_{11} + B_{12}A_{21} & B_{11}A_{12} + B_{12}A_{22} \\ \hline B_{21}A_{11} + B_{22}A_{12} & B_{21}A_{12} + B_{22}A_{22} \\ \hline \end{array} \qquad (3\text{-}17)$$

It will be noted $AB \ne BA$.

A multiplication of the form PBP or $P_mB_{mn}P_n$ is referred to as a quadratic form. The P is a row or column and the B is a rectangular matrix. The **P** may be written as a row or as a column to suit the arrow rule of multiplication discussed earlier. Consider the following problem:

Let
$$P = \begin{array}{c} 1 \\ 2 \end{array} \begin{array}{|c|} \hline P_1 \\ \hline P_2 \\ \hline \end{array}$$

$$B = \begin{array}{c} \\ 1 \\ 2 \end{array} \begin{array}{|c|c|} \hline 1 & 2 \\ \hline B_{11} & B_{12} \\ \hline B_{21} & B_{22} \\ \hline \end{array}$$

It is desired to find **PBP**.

Performing the **PB** operation, we have

$$\begin{array}{|c|c|} \hline P_1 & P_2 \\ \hline \end{array} \quad \begin{array}{|c|c|} \hline B_{11} & B_{12} \\ \hline B_{21} & B_{22} \\ \hline \end{array} = \begin{array}{|c|c|} \hline P_1B_{11}+P_2B_{21} & P_1B_{12}+P_2B_{22} \\ \hline \end{array}$$

Performing the **(PB)P** multiplication, we obtain

$$\begin{array}{|c|c|} \hline P_1B_{11}+P_2B_{21} & P_1B_{12}+P_2B_{22} \\ \hline \end{array} \begin{array}{|c|} \hline P_1 \\ \hline P_2 \\ \hline \end{array} = \begin{array}{l} P_1B_{11}P_1 + P_2B_{21}P_1 \\ + P_1B_{12}P_2 + P_2B_{22}P_2 \end{array} \quad (3\text{-}18)$$

In terms of index notation, we obtain

$$P_m B_{mn} P_n = P_1 B_{1n} P_n + P_2 B_{2n} P_n$$

by letting m assume the values of 1 and 2. Letting $n = 1, 2$, we obtain

$$P_m B_{mn} P_n = P_1 B_{11} P_1 + P_1 B_{12} P_2 + P_2 B_{21} P_1 + P_2 B_{22} P_2 \quad (3\text{-}19)$$

which is the same as the result obtained by the arrow rule.

The transpose of a given matrix A is denoted by A_t and is obtained by interchanging the rows and columns of A. Thus, if

DEVELOPMENT OF TRANSMISSION LOSS FORMULA

$$A = \begin{array}{c} \\ 1 \\ 2 \\ 3 \end{array} \begin{array}{|ccc|} \hline 1 & 2 & 3 \\ \hline 2 & 1 & 3 \\ 7 & 5 & 8 \\ 0 & 6 & 9 \\ \hline \end{array}$$

then

$$A_t = \begin{array}{c} \\ 1 \\ 2 \\ 3 \end{array} \begin{array}{|ccc|} \hline 1 & 2 & 3 \\ \hline 2 & 7 & 0 \\ 1 & 5 & 6 \\ 3 & 8 & 9 \\ \hline \end{array}$$

The unit matrix (1) is defined as a matrix in which all the main diagonal elements are unity and all the off-diagonal elements are zero.

Thus

$$\begin{array}{c} \\ 1 \\ 2 \\ 3 \end{array} \begin{array}{|ccc|} \hline 1 & 2 & 3 \\ \hline 1 & 0 & 0 \\ 0 & 1 & 0 \\ 0 & 0 & 1 \\ \hline \end{array}$$

would be a unit matrix. Usually the zero terms are left blank.

Consider the following multiplication of a matrix A by a unit matrix.

$$\begin{array}{|ccc|} \hline A_{11} & A_{12} & A_{13} \\ A_{21} & A_{22} & A_{23} \\ A_{31} & A_{32} & A_{33} \\ \hline \end{array} \begin{array}{|ccc|} \hline 1 & & \\ & 1 & \\ & & 1 \\ \hline \end{array} = \begin{array}{|ccc|} \hline A_{11} & A_{12} & A_{13} \\ A_{21} & A_{22} & A_{23} \\ A_{31} & A_{32} & A_{33} \\ \hline \end{array}$$

In general, $$AI = A$$
Also $$IA = A \qquad (3\text{–}20)$$

The operation indicated by a matrix C divided by a matrix A is not defined in matrix algebra. Thus, it is not possible to perform an operation of the type indicated by the second equation below:

$$AB = C$$

$$B = \frac{C}{A}$$

To be able to solve for B in the equation above it is necessary to make use of the concept of the inverse of a matrix. The inverse of a matrix A will be denoted by A^{-1} and has the property such that

$$AA^{-1} = 1$$
or
$$A^{-1}A = 1 \qquad (3\text{–}21)$$

Returning to $$AB = C$$

we may multiply both sides of the equation by the inverse of A, thereby obtaining

$$A^{-1}AB = A^{-1}C$$

$$IB = A^{-1}C$$

$$B = A^{-1}C$$

A method of calculation of A^{-1} is given in Section 6.5 of Chapter 6.

3.5 THE CONCEPT OF TRANSFORMATION [3,4]

In our analysis of transmission losses it is necessary that all changes or transformations in the equivalent circuit of Figure 3.7, as denoted by equation 3-6, be made in such a manner that all source powers, the load powers, and transmission losses remain invariant. These transformations may be made through means of transformation matrices which result in logical and systematic steps in the analysis. Also, the use of transformation matrices leads to orderly computational procedures which are very adaptable to calculation on both manual and automatic digital computers.

The concept of the transformation matrix C allows a given circuit to be modified to a new circuit in such a manner that the power input remains invariant. Denote the quantities pertaining to the original circuit by the subscript *old* and the quantities pertaining to the desired new circuit by the subscript *new*. In general, it has been shown by G. Kron [3]

DEVELOPMENT OF TRANSMISSION LOSS FORMULA

that if the set of currents i_old pertaining to the old circuit is related to the new currents i_new by a transformation matrix C such that,

$$i_\text{old} = C i_\text{new} \qquad (3\text{-}22)$$

and if the power is to remain invariant the new set of voltages is given by

$$e_\text{new} = C_t{}^* e_\text{old} \qquad (3\text{-}23)$$

and the new set of impedances is given by

$$Z_\text{new} = C_t{}^* Z_\text{old} C \qquad (3\text{-}24)$$

The matrix $C_t{}^*$ is obtained by conjugating the elements of the matrix C_t. A demonstration of the above relations is given in Appendix 3.1.

This concept is first illustrated in terms of the simple example shown in Figure 3.14 and Figure 3.15. As will be noted from the work already

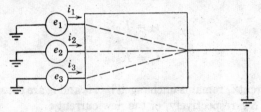

Figure 3.14. Three-source system.

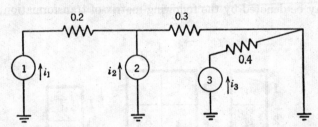

Figure 3.15. Schematic diagram of illustrative three-source system.

presented, this network may be defined by the set of equations which follow:

$$\begin{array}{c|c} 1 & e_1 \\ \hline 2 & e_2 \\ \hline 3 & e_3 \end{array} = \begin{array}{c|ccc} & 1 & 2 & 3 \\ \hline 1 & Z_{11} & Z_{12} & Z_{13} \\ \hline 2 & Z_{21} & Z_{22} & Z_{23} \\ \hline 3 & Z_{31} & Z_{32} & Z_{33} \end{array} \begin{array}{c|c} 1 & i_1 \\ \hline 2 & i_2 \\ \hline 3 & i_3 \end{array} \qquad (3\text{-}25)$$

ECONOMIC OPERATION OF POWER SYSTEMS

$$
\begin{array}{c|ccc}
 & \overset{\rightarrow}{1} & 2 & 3 \\
\hline
1 & 0.5 & 0.3 & 0 \\
= 2 & 0.3 & 0.3 & 0 \\
3 & 0 & 0 & 0.4
\end{array}
\quad
\begin{array}{c|c}
\downarrow & \\
1 & i_1 \\
2 & i_2 \\
3 & i_3
\end{array}
\quad (3\text{-}26)
$$

Let it be required to obtain the new circuit that exists if the old currents are related to the new currents by the relation

$$i_1 = i_1$$

$$i_2 = K_2 i_4 \quad (3\text{-}27)$$

$$i_3 = K_3 i_4$$

where the current i_1 remains unchanged but i_2 and i_3 are constant proportions K_2 and K_3, respectively, of the new current i_4.

The relation between the two sets of currents as given by equation 3-27 may be denoted by the following matrix of transformation:

$$
\begin{array}{c|c}
1 & i_1 \\
2 & i_2 \\
3 & i_3
\end{array}
=
\begin{array}{c|cc}
 & \overset{\rightarrow}{1} & 4 \\
\hline
1 & 1 & \\
2 & & K_2 \\
3 & & K_3
\end{array}
\quad
\begin{array}{c|c}
\downarrow & \\
1 & i_1 \\
4 & i_4
\end{array}
$$

where

$$
C =
\begin{array}{c|cc}
 & 1 & 4 \\
\hline
1 & 1 & \\
2 & & K_2 \\
3 & & K_3
\end{array}
\quad (3\text{-}28)
$$

DEVELOPMENT OF TRANSMISSION LOSS FORMULA

The transpose of C will then be

$$C_t = \begin{array}{c|ccc} & 1 & 2 & 3 \\ \hline 1 & 1 & & \\ \hline 4 & & K_2 & K_3 \end{array} \qquad (3\text{-}29)$$

Since the elements of C_t are real numbers, $C_t^* = C_t$.
The new voltages are given by

$$e_{\text{new}} = C_t e_{\text{old}} = \begin{array}{c|ccc} & 1 & 2 & 3 \\ \hline 1 & 1 & & \\ \hline 4 & & K_2 & K_3 \end{array} \begin{array}{c|c} 1 & e_1 \\ \hline 2 & e_2 \\ \hline 3 & e_3 \end{array}$$

$$= \begin{array}{c|c} 1 & e_1 \\ \hline 2 & K_2 e_2 + K_3 e_3 \end{array} \qquad (3\text{-}30)$$

As indicated previously, the new impedances are given by

$$Z_{\text{new}} = C_t^* Z_{\text{old}} C \qquad (3\text{-}24)$$

The product $Z_{\text{old}} C$ is first calculated as

$$\begin{array}{c|ccc} & 1 & 2 & 3 \\ \hline 1 & Z_{11} & Z_{12} & Z_{13} \\ \hline 2 & Z_{21} & Z_{22} & Z_{23} \\ \hline 3 & Z_{31} & Z_{32} & Z_{33} \end{array} \begin{array}{c|cc} & 1 & 4 \\ \hline 1 & 1 & \\ \hline 2 & & K_2 \\ \hline 3 & & K_3 \end{array} = \begin{array}{c|cc} & 1 & 4 \\ \hline 1 & Z_{11} & Z_{12}K_2 + Z_{13}K_3 \\ \hline 2 & Z_{21} & Z_{22}K_2 + Z_{23}K_3 \\ \hline 3 & Z_{31} & Z_{32}K_2 + Z_{33}K_3 \end{array}$$

In terms of numbers

$$Z_{\text{old}}C = \begin{array}{c|ccc} & 1 & 2 & 3 \\ \hline 1 & 0.5 & 0.3 & \\ 2 & 0.3 & 0.3 & \\ 3 & & & 0.4 \end{array} \quad \begin{array}{c|cc} & 1 & 4 \\ \hline 1 & 1 & \\ 2 & & K_2 \\ 3 & & K_3 \end{array} = \begin{array}{c|cc} & 1 & 4 \\ \hline 1 & 0.5 & 0.3K_2 \\ 2 & 0.3 & 0.3K_2 \\ 3 & & 0.4K_3 \end{array}$$

The operation $C_t{}^*(Z_{\text{old}}C)$ is calculated as

$$\begin{array}{c|ccc} & 1 & 2 & 3 \\ \hline 1 & 1 & & \\ 4 & & K_2 & K_3 \end{array} \quad \begin{array}{c|cc} & 1 & 4 \\ \hline 1 & Z_{11} & Z_{12}K_2 + Z_{13}K_3 \\ 2 & Z_{21} & Z_{22}K_2 + Z_{23}K_3 \\ 3 & Z_{31} & Z_{32}K_2 + Z_{33}K_3 \end{array} = \begin{array}{c|cc} & 1 & 4 \\ \hline 1 & Z_{11} & Z_{14} \\ 4 & Z_{41} & Z_{44} \end{array}$$

where
$$\begin{aligned} Z_{14} &= Z_{12}K_2 + Z_{13}K_3 \\ Z_{41} &= K_2 Z_{21} + K_3 Z_{31} \\ Z_{44} &= K_2 Z_{22} K_2 + K_2 Z_{23} K_3 + K_3 Z_{32} K_2 + K_3 Z_{33} K_3 \end{aligned} \quad (3\text{-}31)$$

In terms of numbers we have

$$\begin{array}{c|ccc} & 1 & 2 & 3 \\ \hline 1 & 1 & & \\ 4 & & K_2 & K_3 \end{array} \quad \begin{array}{c|cc} & 1 & 4 \\ \hline 1 & 0.5 & 0.3K_2 \\ 2 & 0.3 & 0.3K_2 \\ 3 & & 0.4K_3 \end{array} = \begin{array}{c|cc} & 1 & 4 \\ \hline 1 & 0.5 & 0.3K_2 \\ 4 & K_2(0.3) & 0.3K_2{}^2 + 0.4K_3{}^2 \end{array} \quad (3\text{-}32)$$

The new set of equations is

$$\begin{array}{c|c} & \\ \hline 1 & e_1 \\ 4 & e_4 \end{array} = \begin{array}{c|cc} & 1 & 4 \\ \hline 1 & Z_{11} & Z_{14} \\ 4 & Z_{41} & Z_{44} \end{array} \quad \begin{array}{c|c} & \\ \hline 1 & i_1 \\ 4 & i_4 \end{array} \quad (3\text{-}33)$$

DEVELOPMENT OF TRANSMISSION LOSS FORMULA

where the voltages are given by equation 3–30 and the impedances by equation 3–31.

In terms of numbers we have

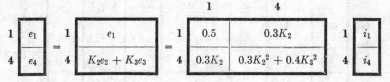

The corresponding new circuit is drawn in Figure 3.16.

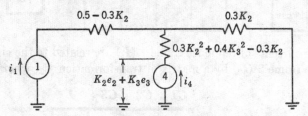

Figure 3.16. Modified circuit after transformation.

3.6 TRANSFORMATION TO REFERENCE FRAME 2

It is desired to eliminate the individual equivalent load currents as variables, since the final result should involve only generator powers. As will be recalled, the equivalent load current at a bus is defined as the sum of the line-charging, synchronous condenser, and load current at that bus. The first assumption involved in the development of a loss formula is the following:

It is assumed that each equivalent load current remains a constant complex fraction of the total equivalent load current.

Define
$$i_L = \sum_j i_{Lj} \qquad (3\text{-}34)$$

By the above assumption $\qquad i_{Lj} = l_j i_L \qquad (3\text{-}35)$

For the system given by Figure 3.10 and equation 3–12 we may write

$$i_{G1} = i_{G1}$$
$$i_{G2} = i_{G2}$$
$$i_{G3} = i_{G3}$$
$$i_{L1} = l_1 i_L$$
$$i_{L2} = l_2 i_L$$

The preceding relation may be written in terms of a matrix of transformation.

$$
\begin{array}{c|c}
G1 & i_{G1} \\ \hline
G2 & i_{G2} \\ \hline
G3 & i_{G3} \\ \hline
L1 & i_{L1} \\ \hline
L2 & i_{L2}
\end{array}
=
\begin{array}{c|cccc}
 & G1 & G2 & G3 & L \\ \hline
G1 & 1 & & & \\ \hline
G2 & & 1 & & \\ \hline
G3 & & & 1 & \\ \hline
L1 & & & & l_1 \\ \hline
L2 & & & & l_2
\end{array}
\begin{array}{c|c}
G1 & i_{G1} \\ \hline
G2 & i_{G2} \\ \hline
G3 & i_{G3} \\ \hline
L & i_L
\end{array}
\qquad (3\text{--}36)
$$

Thus the currents of reference frame 1 (I_1) are related to the currents of reference frame 2 (I_2) by a matrix of transformation $C_2^{\,1}$ where

$$
C_2^{\,1} =
\begin{array}{c|cccc}
 & G1 & G2 & G3 & L \\ \hline
G1 & 1 & & & \\ \hline
G2 & & 1 & & \\ \hline
G3 & & & 1 & \\ \hline
L1 & & & & l_1 \\ \hline
L2 & & & & l_2
\end{array}
\qquad (3\text{--}37)
$$

The symbol $C_k^{\,j}$ is used to indicate the transformation from reference frame or step j to reference frame or step k.

The new impedance matrix is given as indicated by equation 3–24 by $C_t{}^* Z_{\text{old}} C$. Performing the $Z_{\text{old}} C$ operation first, we have

$$
\begin{array}{c|ccccc}
 & G1 & G2 & G3 & L1 & L2 \\ \hline
G1 & Z_{G1\text{--}G1} & Z_{G1\text{--}G2} & Z_{G1\text{--}G3} & Z_{G1\text{--}L1} & Z_{G1\text{--}L2} \\ \hline
G2 & Z_{G2\text{--}G1} & Z_{G2\text{--}G2} & Z_{G2\text{--}G3} & Z_{G2\text{--}L1} & Z_{G2\text{--}L2} \\ \hline
G3 & Z_{G3\text{--}G1} & Z_{G3\text{--}G2} & Z_{G3\text{--}G3} & Z_{G3\text{--}L1} & Z_{G3\text{--}L2} \\ \hline
L1 & Z_{L1\text{--}G1} & Z_{L1\text{--}G2} & Z_{L1\text{--}G3} & Z_{L1\text{--}L1} & Z_{L1\text{--}L2} \\ \hline
L2 & Z_{L2\text{--}G1} & Z_{L2\text{--}G2} & Z_{L2\text{--}G3} & Z_{L2\text{--}L1} & Z_{L2\text{--}L2}
\end{array}
\begin{array}{c|cccc}
 & G1 & G2 & G3 & L \\ \hline
G1 & 1 & & & \\ \hline
G2 & & 1 & & \\ \hline
G3 & & & 1 & \\ \hline
L1 & & & & l_1 \\ \hline
L2 & & & & l_2
\end{array}
$$

DEVELOPMENT OF TRANSMISSION LOSS FORMULA

$$= \begin{array}{c|ccc|c} & G1 & G2 & G3 & L \\ \hline G1 & Z_{G1-G1} & Z_{G1-G2} & Z_{G1-G3} & Z_{G1-L1}l_1 + Z_{G1-L2}l_2 \\ G2 & Z_{G2-G1} & Z_{G2-G2} & Z_{G2-G3} & Z_{G2-L1}l_1 + Z_{G2-L2}l_2 \\ G3 & Z_{G3-G1} & Z_{G3-G2} & Z_{G3-G3} & Z_{G3-L1}l_1 + Z_{G3-L2}l_2 \\ \hline L1 & Z_{L1-G1} & Z_{L1-G2} & Z_{L1-G3} & Z_{L1-L1}l_1 + Z_{L1-L2}l_2 \\ L2 & Z_{L2-G1} & Z_{L2-G2} & Z_{L2-G3} & Z_{L2-L1}l_1 + Z_{L2-L2}l_2 \end{array} \quad (3\text{-}38)$$

Performing the $C_t{}^*(Z_{\text{old}}C)$ operation, we have

$$\begin{array}{c|ccc|cc} & G1 & G2 & G3 & L1 & L2 \\ \hline G1 & 1 & & & & \\ G2 & & 1 & & & \\ G3 & & & 1 & & \\ \hline L & & & & l_1{}^* & l_2{}^* \end{array} \quad \begin{array}{c|ccc|c} & G1 & G2 & G3 & L \\ \hline G1 & Z_{G1-G1} & Z_{G1-G2} & Z_{G1-G3} & Z_{G1-L1}l_1 + Z_{G1-L2}l_2 \\ G2 & Z_{G2-G1} & Z_{G2-G2} & Z_{G2-G3} & Z_{G2-L1}l_1 + Z_{G2-L2}l_2 \\ G3 & Z_{G3-G1} & Z_{G3-G2} & Z_{G3-G3} & Z_{G3-L1}l_1 + Z_{G3-L2}l_2 \\ \hline L1 & Z_{L1-G1} & Z_{L1-G2} & Z_{L1-G3} & Z_{L1-L1}l_1 + Z_{L1-L2}l_2 \\ L2 & Z_{L2-G1} & Z_{L2-G2} & Z_{L2-G3} & Z_{L2-L1}l_1 + Z_{L2-L2}l_2 \end{array}$$

$$= \begin{array}{c|ccc|c} & G1 & G2 & G3 & L \\ \hline G1 & Z_{G1-G1} & Z_{G1-G2} & Z_{G1-G3} & a_1 \\ G2 & Z_{G2-G1} & Z_{G2-G2} & Z_{G2-G3} & a_2 \\ G3 & Z_{G3-G1} & Z_{G3-G2} & Z_{G3-G3} & a_3 \\ \hline L & b_1 & b_2 & b_3 & w \end{array} \quad (3\text{-}39)$$

where $a_1 = Z_{G1-L1}l_1 + Z_{G1-L2}l_2$
$a_2 = Z_{G2-L1}l_1 + Z_{G2-L2}l_2$
$a_3 = Z_{G3-L1}l_1 + Z_{G3-L2}l_2$
$b_1 = l_1{}^*Z_{L1-G1} + l_2{}^*Z_{L2-G1}$ \quad (3-40)
$b_2 = l_1{}^*Z_{L1-G2} + l_2{}^*Z_{L2-G2}$
$b_3 = l_1{}^*Z_{L1-G3} + l_2{}^*Z_{L2-G3}$
$w = l_1{}^*Z_{L1-L1}l_1 + l_1{}^*Z_{L1-L2}l_2 + l_2{}^*Z_{L2-L1}l_1 + l_2{}^*Z_{L2-L2}l_2$

From equation 3–23 it is seen that our new voltages of reference frame 2 are given by $C_t{}^* e_{\text{old}}$ as indicated by

	G1	G2	G3	L1	L2		G1	$E_{G1} - E_R$			G1	$E_{G1} - E_R$
G1	1						G2	$E_{G2} - E_R$			G2	$E_{G2} - E_R$
G2		1					G3	$E_{G3} - E_R$	=		G3	$E_{G3} - E_R$
G3			1				L1	$E_{L1} - E_R$			L	$E_L - E_R$
L				$l_1{}^*$	$l_2{}^*$		L2	$E_{L2} - E_R$				

$$C_{t^*} \qquad\qquad e_{\text{old}} \qquad = \qquad e_{\text{new}}$$

(3–41)

The calculation for $E_L - E_R$ is indicated in detail below.

$$l_1{}^*(E_{L1} - E_R) + l_2{}^*(E_{L2} - E_R) = l_1{}^*E_{L1} + l_2{}^*E_{L2} - (l_1{}^* + l_2{}^*)E_R$$

Define $\qquad\qquad E_L = l_1{}^*E_{L1} + l_2{}^*E_{L2}$

Since $\qquad\qquad (l_1{}^* + l_2{}^*) = 1$

we have $\qquad l_1{}^*(E_{L1} - E_R) + l_2{}^*(E_{L2} - E_R) = E_L - E_R$

From equations 3–36, 3–39, and 3–41 we have

				G1	G2	G3	L			
G1	$E_{G1} - E_R$		G1	Z_{G1-G1}	Z_{G1-G2}	Z_{G1-G3}	a_1		G1	i_{G1}
G2	$E_{G2} - E_R$	=	G2	Z_{G2-G1}	Z_{G2-G2}	Z_{G2-G3}	a_2		G2	i_{G2}
G3	$E_{G3} - E_R$		G3	Z_{G3-G1}	Z_{G3-G2}	Z_{G3-G3}	a_3		G3	i_{G3}
L	$E_L - E_R$		L	b_1	b_2	b_3	w		L	i_L

(3–42)

It will be noted that the effect of each load current has been replaced by a single total load current.

The preceding steps accomplished by the transformation matrix $C_2{}^1$ may be thought of in terms of a number of algebraic steps. Consider the reference frame 1 equations for this example as repeated below:

DEVELOPMENT OF TRANSMISSION LOSS FORMULA

| i_{G1} | i_{G2} | i_{G3} | i_{L1} | i_{L2} |

				G1	G2	G3	L1	L2	
G1	$E_{G1} - E_R$		G1	Z_{G1-G1}	Z_{G1-G2}	Z_{G1-G3}	Z_{G1-L1}	Z_{G1-L2}	
G2	$E_{G2} - E_R$		G2	Z_{G2-G1}	Z_{G2-G2}	Z_{G2-G3}	Z_{G2-L1}	Z_{G2-L2}	
G3	$E_{G3} - E_R$	=	G3	Z_{G3-G1}	Z_{G3-G2}	Z_{G3-G3}	Z_{G3-L1}	Z_{G3-L2}	(3-12)
L1	$E_{L1} - E_R$		L1	Z_{L1-G1}	Z_{L1-G2}	Z_{L1-G3}	Z_{L1-L1}	Z_{L1-L2}	
L2	$E_{L2} - E_R$		L2	Z_{L2-G1}	Z_{L2-G2}	Z_{L2-G3}	Z_{L2-L1}	Z_{L2-L2}	

As before, let

$$i_{L1} = l_1 i_L$$
$$i_{L2} = l_2 i_L \tag{3-43}$$

Substituting equation 3-43 into equation 3-12, we obtain

| i_{G1} | i_{G2} | i_{G3} | i_L |

				G1	G2	G3	L	
G1	$E_{G1} - E_R$		G1	Z_{G1-G1}	Z_{G1-G2}	Z_{G1-G3}	$Z_{G1-L1}l_1 + Z_{G1-L2}l_2$	
G2	$E_{G2} - E_R$		G2	Z_{G2-G1}	Z_{G2-G2}	Z_{G2-G3}	$Z_{G2-L1}l_1 + Z_{G2-L2}l_2$	
G3	$E_{G3} - E_R$	=	G3	Z_{G3-G1}	Z_{G3-G2}	Z_{G3-G3}	$Z_{G3-L1}l_1 + Z_{G3-L2}l_2$	(3-44)
L1	$E_{L1} - E_R$		L1	Z_{L1-G1}	Z_{L1-G2}	Z_{L1-G3}	$Z_{L1-L1}l_1 + Z_{L1-L2}l_2$	
L2	$E_{L2} - E_R$		L2	Z_{L2-G1}	Z_{L2-G2}	Z_{L2-G3}	$Z_{L2-L1}l_1 + Z_{L2-L2}l_2$	

The impedances in equation 3-44 correspond to $\mathbf{Z}_{\text{old}}\mathbf{C}$.

Define a hypothetical load voltage E_L such that the power loss contributed by $(E_{L1} - E_R)i_{L1}^* + (E_{L2} - E_R)i_{L2}^*$ remains invariant. Thus

$$(E_L - E_R)i_L^* = (E_{L1} - E_R)i_{L1}^* + (E_{L2} - E_R)i_{L2}^*$$
$$= (E_{L1} - E_R)l_1^* i_L^* + (E_{L2} - E_R)l_2^* i_L^* \tag{3-45}$$

Dividing equation 3–45 by $i_L{}^*$,

$$(E_L - E_R) = (E_{L1} - E_R)l_1{}^* + (E_{L2} - E_R)l_2{}^* \qquad (3\text{–}46)$$

Performing the operation indicated by equation 3–46 upon equation 3–44, we obtain equation 3–42 as before.

From equation 3–42 we note that the mutual impedances between generators and loads are not symmetrical. That is, $a_1 \neq b_1$, $a_2 \neq b_2$, and $a_3 \neq b_3$. For example,

$$a_1 = Z_{G1-L1}l_1 + Z_{G1-L2}l_2$$

$$= (R_{G1-L1} + jX_{G1-L1})(l_1' + jl_1'')$$

$$\qquad\qquad + (R_{G1-L2} + jX_{G1-L2})(l_2' + jl_2'')$$

$$= R_{G1-L1}l_1' + R_{G1-L2}l_2' - X_{G1-L1}l_1'' - X_{G1-L2}l_2'$$

$$+ j(X_{G1-L1}l_1' + X_{G1-L2}l_2' + R_{G1-L1}l_1'' + R_{G1-L2}l_2'') \qquad (3\text{–}47)$$

$$b_1 = l_1{}^*Z_{L1-G1} + l_2{}^*Z_{L2-G1}$$

$$= (l_1' - jl_1'')(R_{L1-G1} + jX_{L1-G1})$$

$$\qquad + (l_2' - jl_2'')(R_{L2-G1} + jX_{L2-G1})$$

$$= l_1'R_{L1-G1} + l_2'R_{L2-G1} + l_1''X_{L1-G1} + l_2''X_{L2-G1}$$

$$+ j(l_1'X_{L1-G1} + l_2'X_{L2-G1} - l_1''R_{L1-G1} - l_2''R_{L2-G1}) \qquad (3\text{–}48)$$

Consider next the more general case in which the number of sources $= m, n$ and the number of loads $= j, k$ and for which reference frame 1 equations are given by equation 3–6 which is repeated below:

$$\begin{array}{c|c|} & \\ Gm & E_{Gm} - E_R \\ \hline Lj & E_{Lj} - E_R \\ \end{array} = \begin{array}{c|c|c|} & Gn & Lk \\ \hline Gm & Z_{Gm-Gn} & Z_{Gm-Lk} \\ \hline Lj & Z_{Lj-Gn} & Z_{Lj-Lk} \\ \end{array} \begin{array}{c|c|} & \\ Gn & i_{Gn} \\ \hline Lk & i_{Lk} \\ \end{array} \qquad (3\text{–}6)$$

The matrix of transformation $C_2{}^1$ is given by

$$\begin{array}{c|c|} & \\ Gn & i_{Gn} \\ \hline Lk & i_{Lk} \\ \end{array} = \begin{array}{c|c|c|} & Gn & L \\ \hline Gn & 1 & \\ \hline Lk & & l_k \\ \end{array} \begin{array}{c|c|} & \\ Gn & i_{Gn} \\ \hline L & i_L \\ \end{array} \qquad (3\text{–}49)$$

DEVELOPMENT OF TRANSMISSION LOSS FORMULA

$$C_2^1 = \begin{array}{c|c|c} & Gn & L \\ \hline Gn & 1 & \\ \hline Lk & & l_k \end{array} \qquad (3\text{-}50)$$

The matrix $Lk\ \boxed{l_k}$ (with row label L) is a column matrix with the number of elements equal to k, the number of load currents. This follows by inspection of the matrix, since the number of columns correspond to L and the number of rows to Lk. Since there is only one hypothetical load current, there is only one column. The transpose of C_2^1 is given by

$$(C_2^1)_t = \begin{array}{c|c|c} & Gm & Lj \\ \hline Gm & 1 & \\ \hline L & & l_j \end{array}$$

The matrix $L\ \boxed{l_j}$ (with column label Lj) is a row matrix with the same elements as the column matrix $Lk\ \boxed{l_k}$ but with the numbers written in a row instead of a column. The resulting voltages, impedances, and currents are given by

$$\begin{array}{c|c} Gm & E_{Gm} - E_R \\ \hline L & E_L - E_R \end{array} = \begin{array}{c|c|c} & Gn & L \\ \hline Gm & Z_{Gm-Gn} & a_m \\ \hline L & b_n & w \end{array} \begin{array}{c|c} & Gn \\ \hline Gn & i_{Gn} \\ \hline L & i_L \end{array} \qquad (3\text{-}51)$$

where

$$\begin{aligned} a_m &= Z_{Gm-Lk} l_k \\ b_n &= l_j^* Z_{Lj-Gn} \\ w &= l_j^* Z_{Lj-Lk} l_k \\ E_L &= l_j^* E_{Lj} \end{aligned} \qquad (3\text{-}52)$$

By means of the above transformation the circuit of Figure 3.7 has been changed to the circuit given in Figure 3.17. The load point L does

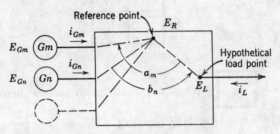

Figure 3.17. Reference frame 2.

not exist in the actual network, and consequently it is referred to as a hypothetical load point. As previously noted, the mutual impedances between the generators and loads are not equal. As noted by equation 3–51 and Figure 3.17, the component of the voltage drop $E_{Gm} - E_R$ due to load current i_L is given by $a_m i_L$. Similarly, the component of the voltage drop $E_L - E_R$ due to current i_{Gn} is given by $b_n i_{Gn}$. The impedance w is the self impedance existing between the hypothetical load point and the reference bus.

3.7 TRANSFORMATION TO REFERENCE FRAME 3

As will be noted in equations 3–42 and 3–51 and also Figure 3.17, the individual load currents have been eliminated as variables and replaced by the total load current i_L. The next step in our analysis involves eliminating the total load current i_L as a variable. We may accomplish this by the relationship that the summation of the source currents must be equal and opposite to the summation of the load currents. Thus

$$\sum_n i_{Gn} = -i_L$$

For the system of Figure 3.10 and equation 3–42 we may write

$$i_{G1} = i_{G1}$$
$$i_{G2} = i_{G2}$$
$$i_{G3} = i_{G3}$$
$$i_L = -(i_{G1} + i_{G2} + i_{G3})$$

(3–53)

DEVELOPMENT OF TRANSMISSION LOSS FORMULA

The above relation may be written in terms of a matrix of transformation as indicated below:

$$\begin{array}{c|c} & \\ G1 & i_{G1} \\ G2 & i_{G2} \\ G3 & i_{G3} \\ \hline L & i_L \end{array} = \begin{array}{c|ccc} & G1 & G2 & G3 \\ \hline G1 & 1 & & \\ G2 & & 1 & \\ G3 & & & 1 \\ \hline L & -1 & -1 & -1 \end{array} \begin{array}{c|c} G1 & i_{G1} \\ G2 & i_{G2} \\ G3 & i_{G3} \end{array} \quad (3\text{-}54)$$

Thus the currents of reference frame 2 (I_2) are related to the currents of reference frame 3 (I_3) by a matrix of transformation $C_3{}^2$ where

$$C_3{}^2 = \begin{array}{c|ccc} & G1 & G2 & G3 \\ \hline G1 & 1 & & \\ G2 & & 1 & \\ G3 & & & 1 \\ \hline L & -1 & -1 & -1 \end{array} \quad (3\text{-}55)$$

The new voltages of reference frame 3 are given by

$$\begin{array}{c|cccc} & G1 & G2 & G3 & L \\ \hline G1 & 1 & & & -1 \\ G2 & & 1 & & -1 \\ G3 & & & 1 & -1 \end{array} \begin{array}{c|c} G1 & E_{G1} - E_R \\ G2 & E_{G2} - E_R \\ G3 & E_{G3} - E_R \\ L & E_L - E_R \end{array} = \begin{array}{c|c} G1 & E_{G1} - E_L \\ G2 & E_{G2} - E_L \\ G3 & E_{G3} - E_L \end{array} \quad (3\text{-}56)$$

$$C_t{}^* \qquad\qquad e_{\text{old}} \qquad = \qquad e_{\text{new}}$$

76 ECONOMIC OPERATION OF POWER SYSTEMS

The new impedance matrix, as indicated by equation 3-24, is given by $C_t{}^*Z_{\text{old}}C$. Performing the $C_t{}^*Z_{\text{old}}$ operation first, we have

	G1	G2	G3	L
G1	1			−1
G2		1		−1
G3			1	−1

	G1	G2	G3	L
G1	Z_{G1-G1}	Z_{G1-G2}	Z_{G1-G3}	a_1
G2	Z_{G2-G1}	Z_{G2-G2}	Z_{G2-G3}	a_2
G3	Z_{G3-G1}	Z_{G3-G2}	Z_{G3-G3}	a_3
L	b_1	b_2	b_3	w

$$= \begin{array}{c|cccc} & G1 & G2 & G3 & L \\ \hline G1 & Z_{G1-G1}-b_1 & Z_{G1-G2}-b_2 & Z_{G1-G3}-b_3 & a_1-w \\ G2 & Z_{G2-G1}-b_1 & Z_{G2-G2}-b_2 & Z_{G2-G3}-b_3 & a_2-w \\ G3 & Z_{G3-G1}-b_1 & Z_{G3-G2}-b_2 & Z_{G3-G3}-b_3 & a_3-w \end{array} \quad (3\text{-}57)$$

Performing the $(C_t{}^*Z_{\text{old}})C$ operation, we have

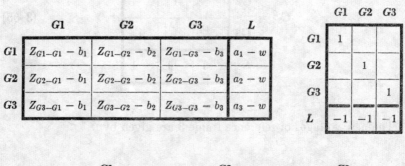

$$= \begin{array}{c|ccc} & G1 & G2 & G3 \\ \hline G1 & Z_{G1-G1}-b_1-a_1+w & Z_{G1-G2}-b_2-a_1+w & Z_{G1-G3}-b_3-a_1+w \\ G2 & Z_{G2-G1}-b_1-a_2+w & Z_{G2-G2}-b_2-a_2+w & Z_{G2-G3}-b_3-a_2+w \\ G3 & Z_{G3-G1}-b_1-a_3+w & Z_{G3-G2}-b_2-a_3+w & Z_{G3-G3}-b_3-a_3+w \end{array}$$

(3-58)

DEVELOPMENT OF TRANSMISSION LOSS FORMULA

From equations 3–54, 3–56, and 3–58, the reference frame 3 currents, impedances, and voltages are given by

$$
\begin{array}{c|c}
G1 & E_{G1} - E_L \\ \hline
G2 & E_{G2} - E_L \\ \hline
G3 & E_{G3} - E_L
\end{array} =
$$

| | i_{G1} | i_{G2} | i_{G3} |

		G1	G2	G3
	G1	$Z_{G1-G1} - a_1 - b_1 + w$	$Z_{G1-G2} - a_1 - b_2 + w$	$Z_{G1-G3} - a_1 - b_3 + w$
=	G2	$Z_{G2-G1} - a_2 - b_1 + w$	$Z_{G2-G2} - a_2 - b_2 + w$	$Z_{G2-G3} - a_2 - b_3 + w$
	G3	$Z_{G3-G1} - a_3 - b_1 + w$	$Z_{G3-G2} - a_3 - b_2 + w$	$Z_{G3-G3} - a_3 - b_3 + w$

(3–59)

It will be noted from equation 3–59 that only the generator currents appear as variables.

Also

$$
\begin{array}{c|c}
G1 & E_{G1} - E_L \\ \hline
G2 & E_{G2} - E_L \\ \hline
G3 & E_{G3} - E_L
\end{array}
=
\begin{array}{c|ccc}
 & G1 & G2 & G3 \\ \hline
G1 & Z_{1\text{-}1} & Z_{1\text{-}2} & Z_{1\text{-}3} \\ \hline
G2 & Z_{2\text{-}1} & Z_{2\text{-}2} & Z_{2\text{-}3} \\ \hline
G3 & Z_{3\text{-}1} & Z_{3\text{-}2} & Z_{3\text{-}3}
\end{array}
\begin{array}{c|c}
G1 & i_{G1} \\ \hline
G2 & i_{G2} \\ \hline
G3 & i_{G3}
\end{array}
\quad (3\text{–}60)
$$

where $Z_{m-n} = Z_{Gm-Gn} - a_m - b_n + w$ (3–61)

The results of performing the operations indicated by equations 3–22, 3–23, and 3–24 with transformation $C_3{}^2$ may be visualized by a number of algebraic steps. Consider the reference frame 2 equations for this example as shown in equation 3–42:

ECONOMIC OPERATION OF POWER SYSTEMS

	i_{G1}	i_{G2}	i_{G3}	i_L

			G1	G2	G3	L
G1	$E_{G1}-E_R$	G1	Z_{G1-G1}	Z_{G1-G2}	Z_{G1-G3}	a_1
G2	$E_{G2}-E_R$	G2	Z_{G2-G1}	Z_{G2-G2}	Z_{G2-G3}	a_2
G3	$E_{G3}-E_R$	G3	Z_{G3-G1}	Z_{G3-G2}	Z_{G3-G3}	a_3
L	E_L-E_R	L	b_1	b_2	b_3	w

$$\text{(3-42)}$$

Subtracting $\quad E_L - E_R = b_1 i_{G1} + b_2 i_{G2} + b_3 i_{G3} + i_L w$

from each of the previous equations in 3–42 we have

	i_{G1}	i_{G2}	i_{G3}	i_L

			G1	G2	G3	L
G1	$E_{G1}-E_L$	G1	$Z_{G1-G1}-b_1$	$Z_{G1-G2}-b_2$	$Z_{G1-G3}-b_3$	$a_1 - w$
G2	$E_{G2}-E_L$	G2	$Z_{G2-G1}-b_1$	$Z_{G2-G2}-b_2$	$Z_{G2-G3}-b_3$	$a_2 - w$
G3	$E_{G3}-E_L$	G3	$Z_{G3-G1}-b_1$	$Z_{G3-G2}-b_2$	$Z_{G3-G3}-b_3$	$a_3 - w$

$$\text{(3-62)}$$

The voltages in equation 3–62 correspond to $C_t^* e_{\text{old}}$, as indicated by equation 3–56. Also, the impedances indicated by equation 3–62 correspond to $C_t^* Z_{\text{old}}$, as indicated by equation 3–57.

Substituting $\quad i_L = -(i_{G1} + i_{G2} + i_{G3})$

into equation 3–62, we obtain equation 3–59 as before.

From equations 3–59, 3–60, and 3–61 we note, as in reference frame 2, that the mutual impedances are not symmetrical.

DEVELOPMENT OF TRANSMISSION LOSS FORMULA

Thus $\quad Z_{1-2} = Z_{G1-G2} - a_1 - b_2 + w = R_{1-2} + jX_{1-2} \quad (3\text{-}63)$

$$Z_{2-1} = Z_{G2-G1} - a_2 - b_1 + w = R_{2-1} + jX_{2-1} \quad (3\text{-}64)$$

From equations 3-40, 3-47, and 3-48,

$$R_{1-2} = R_{G1-G2} - (R_{G1-L1}l_1' + R_{G1-L2}l_2' - X_{G1-L1}l_1''$$
$$- X_{G1-L2}l_2'') - (R_{L1-G2}l_1' + R_{L2-G2}l_2' + X_{L1-G2}l_1''$$
$$+ X_{L2-G2}l_2'') + w' \quad (3\text{-}65)$$

$$X_{1-2} = X_{G1-G2} - (X_{G1-L1}l_1' + X_{G1-L2}l_2' + R_{G1-L1}l_1''$$
$$+ R_{G1-L2}l_2'') - (X_{L1-G2}l_1' + X_{L2-G2}l_2' - R_{L1-G2}l_1''$$
$$- R_{L2-G2}l_2'') + w'' \quad (3\text{-}66)$$

$$R_{2-1} = R_{G2-G1} - (R_{G2-L1}l_1' + R_{G2-L2}l_2' - X_{G2-L1}l_1''$$
$$- X_{G2-L2}l_2'') - (l_1'R_{L1-G1} + l_2'R_{L2-G1} + l_1''X_{L1-G1}$$
$$+ l_2''X_{L2-G1}) + w' \quad (3\text{-}67)$$

$$X_{2-1} = X_{G2-G1} - (X_{G2-L1}l_1' + X_{G2-L2}l_2' + R_{G2-L1}l_1''$$
$$+ R_{G2-L2}l_2'') - (l_1'X_{L1-G1} + l_2'X_{L2-G1} - l_1''R_{L1-G1}$$
$$- l_2''P_{L2-G1}) + w'' \quad (3\text{-}68)$$

where $\quad w = w' + jw'' \quad (3\text{-}69)$

It will be noted that

$$R_{1-2} - R_{2-1} = 2(X_{G1-L1}l_1'' + X_{G1-L2}l_2'' - X_{L1-G2}l_1'' - X_{L2-G2}l_2'')$$
$$(3\text{-}70)$$

$$X_{1-2} - X_{2-1} = 2(-R_{G1-L2}l_1'' - R_{G1-L2}l_2'' + R_{L1-G2}l_1'' + R_{L2-G2}l_2'')$$
$$(3\text{-}71)$$

The asymmetry in the real part of $Z_{m\text{-}n}$ results from terms involving the products of imaginary load currents and mutual reactances between generators and loads. The asymmetry in the imaginary part of $Z_{m\text{-}n}$ results from terms involving the products of imaginary load currents and mutual resistances between generators and load.

The reference frame 2 equations for the general case are given by equation 3-51:

80 ECONOMIC OPERATION OF POWER SYSTEMS

$$\begin{array}{c|c} Gm & E_{Gm} - E_R \\ \hline L & E_L - E_R \end{array} = \begin{array}{c|cc} & Gn & L \\ Gn & Z_{Gm-Gn} & a_m \\ \hline L & b_n & w \end{array} \begin{array}{c|c} Gn & i_{Gn} \\ \hline L & i_L \end{array} \qquad (3\text{-}51)$$

The matrix of transformation $C_3{}^2$ is given by

$$C_3{}^2 = \begin{array}{c|c} & Gn \\ Gn & 1 \\ \hline L & t_n \end{array} \qquad (3\text{-}72)$$

where $t_n = -1$ for all values of n.

By application of equations 3–22, 3–23, and 3–24, we obtain

$$\boxed{E_{Gm} - E_L} = \boxed{Z_{Gm-Gn} - a_m - b_n + w} \quad \boxed{i_{Gn}} \qquad (3\text{-}73)$$

$$= \boxed{Z_{m-n}} \quad \boxed{i_{Gn}}$$

The circuit of reference frame 2, given by Figure 3.17, has been modified as indicated by equations 3–59 and 3–73 to that given in Figure 3.18.

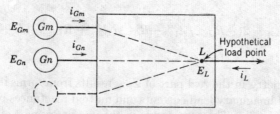

Figure 3.18. Reference frame 3.

As noted by 3–70 and 3–71, the mutual impedances are not symmetrical. Consequently, it is not possible to represent this equivalent circuit on the network analyzer through the use of static circuit ele-

DEVELOPMENT OF TRANSMISSION LOSS FORMULA

ments. The losses in the equivalent circuit of Figure 3.18 correspond to the losses in the transmission lines of the original circuit.

3.8 CALCULATION OF LOSSES

The real losses in the equivalent circuit of Figure 3.18 and equation 3-73 may be calculated as follows:

$$P_L = \mathcal{R} I_3^* E_3 \qquad (3\text{-}74)$$

$$= \mathcal{R} I_3^* Z_3 I_3 \qquad (3\text{-}75)$$

where E_3, I_3, and Z_3 denote reference frame 3 quantities and the symbol $\mathcal{R}$ denotes the real part of $I_3^* E_3$.

Let us define the real and imaginary components of i_{Gn} by i_{dn} and i_{qn}, respectively.

Thus
$$i_{Gn} = i_{dn} + ji_{qn} \qquad (3\text{-}76)$$

For the system of equation 3-60 we have

$$\begin{array}{c}G1\\G2\\G3\end{array}\begin{vmatrix}E_{G1} - E_L\\E_{G2} - E_L\\E_{G3} - E_L\end{vmatrix} = \begin{array}{c}G1\\G2\\G3\end{array}\begin{vmatrix}Z_{1-1} & Z_{1-2} & Z_{1-3}\\Z_{2-1} & Z_{2-2} & Z_{2-3}\\Z_{3-1} & Z_{3-2} & Z_{3-3}\end{vmatrix}\begin{array}{c}G1\\G2\\G3\end{array}\begin{vmatrix}i_{d1} + ji_{q1}\\i_{d2} + ji_{q2}\\i_{d3} + ji_{q3}\end{vmatrix} \qquad (3\text{-}77)$$

$$E_3 \qquad = \qquad Z_3 \qquad\qquad I_3$$

where $\quad Z_{m-n} = Z_{Gm-Gn} - a_m - b_n + w$

Then

$$Z_3 I_3 = \begin{array}{c}G1\\ \\ \\G2\\ \\ \\G3\end{array}\begin{vmatrix}(R_{1-1}i_{d1} + R_{1-2}i_{d2} + R_{1-3}i_{d3} - X_{1-1}i_{q1} - X_{1-2}i_{q2}\\ - X_{1-3}i_{q3}) + j(R_{1-1}i_{q1} + R_{1-2}i_{q2} + R_{1-3}i_{q3}\\ + X_{1-1}i_{d1} + X_{1-2}i_{d2} + X_{1-3}i_{d3})\\\hline(R_{2-1}i_{d1} + R_{2-2}i_{d2} + R_{2-3}i_{d3} - X_{2-1}i_{q1} - X_{2-2}i_{q2}\\ - X_{2-3}i_{q3}) + j(R_{2-1}i_{q1} + R_{2-2}i_{q2} + R_{2-3}i_{q3}\\ + X_{2-1}i_{d1} + X_{2-2}i_{d2} + X_{2-3}i_{d3})\\\hline(R_{3-1}i_{d1} + R_{3-2}i_{d2} + R_{3-3}i_{d3} - X_{3-1}i_{q1} - X_{3-2}i_{q2}\\ - X_{3-3}i_{q3}) + j(R_{3-1}i_{q1} + R_{3-2}i_{q2} + R_{3-3}i_{q3}\\ + X_{3-1}i_{d1} + X_{3-2}i_{d2} + X_{3-3}i_{d3})\end{vmatrix}$$

82 ECONOMIC OPERATION OF POWER SYSTEMS

and

$$
\begin{aligned}
\Re I_3{}^* Z_3 I_3 = {} & i_{d1}R_{1-1}i_{d1} + i_{d1}R_{1-2}i_{d2} + i_{d1}R_{1-3}i_{d3} - i_{d1}X_{1-1}i_{q1} \\
& - i_{d1}X_{1-2}i_{q2} - i_{d1}X_{1-3}i_{q3} + i_{q1}R_{1-1}i_{q1} \\
& + i_{q1}R_{1-2}i_{q2} + i_{q1}R_{1-3}i_{q3} + i_{q1}X_{1-1}i_{d1} \\
& + i_{q1}X_{1-2}i_{d2} + i_{q1}X_{1-3}i_{d3} + i_{d2}R_{2-1}i_{d1} \\
& + i_{d2}R_{2-2}i_{d2} + i_{d2}R_{2-3}i_{d3} - i_{d2}X_{2-1}i_{q1} \\
& - i_{d2}X_{2-2}i_{q2} - i_{d2}X_{2-3}i_{q3} + i_{q2}R_{2-1}i_{q1} \\
& + i_{q2}R_{2-2}i_{q2} + i_{q2}R_{2-3}i_{q3} + i_{q2}X_{2-1}i_{d1} \\
& + i_{q2}X_{2-2}i_{d2} + i_{q2}X_{2-3}i_{d3} + i_{d3}R_{3-1}i_{d1} \\
& + i_{d3}R_{3-2}i_{d2} + i_{d3}R_{3-3}i_{d3} - i_{d3}X_{3-1}i_{q1} \\
& - i_{d3}X_{3-2}i_{q2} - i_{d3}X_{3-3}i_{q3} + i_{q3}R_{3-1}i_{q1} \\
& + i_{q3}R_{3-2}i_{q2} + i_{q3}R_{3-3}i_{q3} + i_{q3}X_{3-1}i_{d1} \\
& + i_{q3}X_{3-2}i_{d2} + i_{q3}X_{3-3}i_{d3}
\end{aligned}
\quad (3\text{-}78)
$$

Combining terms, we have

$$
\begin{aligned}
\Re I_3{}^* Z_3 I_3 = {} & i_{d1}R_{1-1}i_{d1} + 2i_{d1}\left(\frac{R_{1-2} + R_{2-1}}{2}\right)i_{d2} + 2i_{d1}\left(\frac{R_{1-3} + R_{3-1}}{2}\right)i_{d3} \\
& + i_{d2}R_{2-2}i_{d2} + 2i_{d2}\left(\frac{R_{2-3} + R_{3-2}}{2}\right)i_{d3} + i_{d3}(R_{3-3})i_{d3} \\
& + i_{q1}R_{1-1}i_{q1} + 2i_{q1}\left(\frac{R_{1-2} + R_{2-1}}{2}\right)i_{q2} + 2i_{q1}\left(\frac{R_{1-3} + R_{3-1}}{2}\right)i_{q3} \\
& + i_{q2}R_{2-2}i_{q2} + 2i_{q2}\left(\frac{R_{2-3} + R_{3-2}}{2}\right)i_{q3} + i_{q3}(R_{3-3})i_{q3} \\
& - 2i_{d1}\left(\frac{+X_{1-2} - X_{2-1}}{2}\right)i_{q2} - 2i_{d1}\left(\frac{+X_{1-3} - X_{3-1}}{2}\right)i_{q3} \\
& - 2i_{d2}\left(\frac{+X_{2-1} - X_{1-2}}{2}\right)i_{q1} \\
& - 2i_{d2}\left(\frac{+X_{2-3} - X_{3-2}}{2}\right)i_{q3} - 2i_{d3}\left(\frac{+X_{3-1} - X_{1-3}}{2}\right)i_{q1} \\
& - 2i_{d3}\left(\frac{+X_{3-2} - X_{2-3}}{2}\right)i_{q2}
\end{aligned}
\quad (3\text{-}79)
$$

It will be noted in the foregoing that only the symmetric part of the frame 3 resistances and skew-symmetric part of the frame 3 reactances contribute to a real loss.

A matrix is called symmetric if the same components occur on both sides of the main diagonal line. The main diagonal line is a line drawn from the upper left-hand corner of a matrix to the lower right hand corner. The following is an example of a symmetric matrix:

$$F = \begin{array}{c|ccc} & 1 & 2 & 3 \\ \hline 1 & 3 & 1 & 2 \\ 2 & 1 & 7 & 4 \\ 3 & 2 & 4 & 8 \end{array}$$

It will be recalled from the definition of a transpose given in Section 3.4 that the transpose of a symmetric matrix is identical with the original matrix.

A matrix is denoted as skew-symmetric if the components along the main diagonal are zero and if the same components occur on both sides of the main diagonal line but with opposite signs.

The following is an example of a skew-symmetric matrix:

$$H = \begin{array}{c|ccc} & 1 & 2 & 3 \\ \hline 1 & 0 & +4 & +7 \\ 2 & -4 & 0 & +3 \\ 3 & -7 & -3 & 0 \end{array}$$

In general, a given matrix A may be divided into a symmetric matrix B and a skew-symmetric matrix C such that

$$A = B + C$$

where the symmetric part is given by

$$B = \frac{A + A_t}{2} \qquad (3\text{-}80)$$

and the skew symmetric part is given by

$$C = \frac{A - A_t}{2} \qquad (3\text{-}81)$$

In terms of index notation we write

$$A_{jk} = B_{jk} + C_{jk}$$

where the symmetric part is given by

$$B_{jk} = \frac{A_{jk} + A_{kj}}{2}$$

and the skew-symmetric part is given by

$$C_{jk} = \frac{A_{jk} - A_{kj}}{2}$$

Thus, if

$$A = \begin{array}{c|ccc} & 1 & 2 & 3 \\ \hline 1 & 3 & 1 & 5 \\ 2 & 4 & 2 & 9 \\ 3 & 6 & 8 & 5 \end{array}$$

then
B = symmetric part of A

$$= \begin{array}{c|ccc} & 1 & 2 & 3 \\ \hline 1 & \frac{3+3}{2} & \frac{1+4}{2} & \frac{5+6}{2} \\ 2 & \frac{4+1}{2} & \frac{2+2}{2} & \frac{9+8}{2} \\ 3 & \frac{6+5}{2} & \frac{8+9}{2} & \frac{5+5}{2} \end{array} = \begin{array}{c|ccc} & 1 & 2 & 3 \\ \hline 1 & 3 & 2.5 & 5.5 \\ 2 & 2.5 & 2 & 8.5 \\ 3 & 5.5 & 8.5 & 5 \end{array}$$

and
C = skew-symmetric part of A

$$= \begin{array}{c|ccc} & 1 & 2 & 3 \\ \hline 1 & \frac{3-3}{2} & \frac{1-4}{2} & \frac{5-6}{2} \\ 2 & \frac{4-1}{2} & \frac{2-2}{2} & \frac{9-8}{2} \\ 3 & \frac{6-5}{2} & \frac{8-9}{2} & \frac{5-5}{2} \end{array} = \begin{array}{c|ccc} & 1 & 2 & 3 \\ \hline 1 & 0 & -1.5 & -0.5 \\ 2 & +1.5 & 0 & +0.5 \\ 3 & +0.5 & -0.5 & 0 \end{array}$$

DEVELOPMENT OF TRANSMISSION LOSS FORMULA

In the case of a quadratic form, such as $\boldsymbol{PAP}$ or $P_j A_{jk} P_k$, in which the elements of $\boldsymbol{P}$ and $\boldsymbol{A}$ are real numbers, the matrix $\boldsymbol{A}$ may always be replaced by the symmetric part $(A + A_t)/2$, since the components resulting from the skew-symmetric part reduce to zero.

That is,
$$P\left(\frac{A - A_t}{2}\right)P = 0$$

For example, consider the expression $P_j A_{jk} P_k$ for $j, k = 1, 2, 3$:

$$P_j A_j P_k = P_1 A_{1k} P_k$$
$$+ P_2 A_{2k} P_k$$
$$+ P_3 A_{3k} P_k$$
$$= P_1 A_{11} P_1 + P_1 A_{12} P_2 + P_1 A_{13} P_3$$
$$+ P_2 A_{21} P_1 + P_2 A_{22} P_2 + P_2 A_{23} P_3$$
$$+ P_3 A_{31} P_1 + P_3 A_{32} P_2 + P_3 A_{33} P_3$$

The result obtained using the symmetric part of A_{jk} is

$$P_j B_{jk} P_k =$$
$$P_1\left(\frac{A_{11} + A_{11}}{2}\right)P_1 + P_1\left(\frac{A_{12} + A_{21}}{2}\right)P_2 + P_1\left(\frac{A_{13} + A_{31}}{2}\right)P_3$$
$$+ P_2\left(\frac{A_{21} + A_{12}}{2}\right)P_1 + P_2\left(\frac{A_{22} + A_{22}}{2}\right)P_2 + P_2\left(\frac{A_{23} + A_{32}}{2}\right)P_3$$
$$+ P_3\left(\frac{A_{31} + A_{13}}{2}\right)P_1 + P_3\left(\frac{A_{32} + A_{23}}{2}\right)P_2 + P_3\left(\frac{A_{33} + A_{33}}{2}\right)P_3$$

It is noted that $\quad P_j A_{jk} P_k = P_j B_{jk} P_k$

Next consider the skew-symmetric part of A_{jk}. We then have

$$P_j C_{jk} P_k =$$
$$P_1\left(\frac{A_{11} - A_{11}}{2}\right)P_1 + P_1\left(\frac{A_{12} - A_{21}}{2}\right)P_2 + P_1\left(\frac{A_{13} - A_{31}}{2}\right)P_3$$
$$+ P_2\left(\frac{A_{21} - A_{12}}{2}\right)P_1 + P_2\left(\frac{A_{22} - A_{22}}{2}\right)P_2 + P_2\left(\frac{A_{23} - A_{32}}{2}\right)P_3$$
$$+ P_3\left(\frac{A_{31} - A_{13}}{2}\right)P_1 + P_3\left(\frac{A_{32} - A_{23}}{2}\right)P_1 + P_3\left(\frac{A_{33} - A_{33}}{2}\right)P_3$$
$$= 0$$

86   ECONOMIC OPERATION OF POWER SYSTEMS

Returning to the general case in which the number of sources $= m, n$ we have

$$Z_3 I_3 = Z_{m-n}(i_{dn} + j i_{qn})$$
$$= (R_{m-n} i_{dn} - X_{m-n} i_{qn}) + j(R_{m-n} i_{qn} + X_{m-n} i_{dn})$$

$$\Re I_3^* Z_3 I_3 = \Re(i_{dm} - j i_{qm}) Z_{m-n}(i_{dn} + j i_{qn})$$
$$= i_{dm} R_{m-n} i_{dn} - i_{dm} X_{m-n} i_{qn} + i_{qm} R_{m-n} i_{qn} + i_{qm} X_{m-n} i_{dn}$$
$$= i_{dm} R_{m-n} i_{dn} + i_{qm} R_{m-n} i_{qn} - i_{dm} X_{m-n} i_{qn} + i_{dm} X_{n-m} i_{qn}$$
$$= i_{dm} R_{m-n} i_{dn} + i_{qm} R_{m-n} i_{qn}$$
$$\quad - 2 i_{dm} \frac{(+X_{m-n} - X_{n-m})}{2} i_{qn} \tag{3-82}$$

$$= i_{dm} \left[\frac{R_{m-n} + R_{n-m}}{2}\right] i_{dn} - 2 i_{dm} \left[\frac{+X_{m-n} - X_{n-m}}{2}\right] i_{qn}$$
$$\quad + i_{qm} \left[\frac{R_{m-n} + R_{n-m}}{2}\right] i_{qn} \tag{3-83}$$

If $m, n = 1, 2, 3$, equation 3–83 reduces to equation 3–79. From equation 3–73

$$Z_{m-n} = Z_{Gm-Gn} - a_m - b_n + w$$

The real symmetrical part of Z_{m-n} is given by

$$\frac{R_{m-n} + R_{n-m}}{2}$$
$$= \Re \tfrac{1}{2}[Z_{m-n} + Z_{n-m}]$$
$$= \Re \tfrac{1}{2}[Z_{Gm-Gn} - a_m - b_n + w + Z_{Gn-Gm} - a_n - b_m + w]$$
$$= \Re \tfrac{1}{2}[Z_{Gm-Gn} + Z_{Gn-Gm} - (a_m + b_m) - (a_n + b_n) + 2w]$$
$$= R_{Gm-Gn} - d_m - d_n + w' \tag{3-84}$$

where
$$d_m = \Re\left(\frac{a_m + b_m}{2}\right) \tag{3-85}$$

$$w' = \Re w \tag{3-86}$$

The calculation of d_m is first discussed in terms of the system of equation 3–60. It will be recalled from equations 3–40, 3–47, and 3–48 that

$$a_1 = Z_{G1-L1} l_1 + Z_{G1-L2} l_2$$
$$b_1 = Z_{L1-G1} l_1^* + Z_{L2-G1} l_2^*$$

DEVELOPMENT OF TRANSMISSION LOSS FORMULA

Then $\quad a_1 + b_1 = Z_{G1-L1}l_1 + Z_{L1-G1}l_1{}^* + Z_{G1-L2}l_2 + Z_{L2-G1}l_2{}^*$

but $\quad Z_{G1-L1} = Z_{L1-G1} \quad Z_{G1-L2} = Z_{L2-G1}$

Hence $\quad a_1 + b_1 = Z_{L1-G1}(l_1 + l_1{}^*) + Z_{L2-G1}(l_2 + l_2{}^*)$

Let $\quad l_1 = l_1' + jl_1'' \quad l_2 = l_2' + jl_2''$

Then $\quad a_1 + b_1 = 2Z_{L1-G1}(l_1') + 2Z_{L2-G1}(l_2')$

and $\quad \mathcal{R}\dfrac{a_1 + b_1}{2} = d_1 = R_{L1-G1}l_1' + R_{L2-G1}l_2' \quad\quad (3\text{-}87)$

Similarly,

$$d_2 = \mathcal{R}\frac{a_2 + b_2}{2} = R_{L1-G2}l_1' + R_{L2-G2}l_2' \quad\quad (3\text{-}88)$$

Thus

$$\frac{R_{1-2} + R_{2-1}}{2} = R_{G1-G2} - d_1 - d_2 + w'$$

$$= R_{G1-G2} - R_{L1-G1}l_1' - R_{L2-G1}l_2'$$

$$\phantom{= R_{G1-G2}} - R_{L1-G2}l_1' - R_{L2-G2}l_2' + w' \quad\quad (3\text{-}89)$$

This result may also be obtained by taking one half the sum of equations 3–65 and 3–67.

For our general system it will be recalled from equation 3–52 that

$$a_m = Z_{Gm-Lk}l_k$$

$$b_n = l_j{}^*Z_{Lj-Gn}$$

Since the indices n, m and j, k are dummy indices, we may write

$$b_m = l_k{}^*Z_{Lk-Gm} = Z_{Lk-Gm}l_k{}^*$$

Then $\quad a_m + b_m = Z_{Gm-Lk}l_k + Z_{Lk-Gm}l_k{}^*$

Since $\quad Z_{Gm-Lk} = Z_{Lk-Gm}$

we have $\quad a_m + b_m = Z_{Lk-Gm}(l_k + l_k{}^*)$

$$ = 2Z_{Lk-Gm}l_k'$$

where $\quad l_k' = \mathcal{R}l_k$

Then $\quad \mathcal{R}\dfrac{a_m + b_m}{2} = d_m = R_{Lk-Gm}l_k' \quad\quad (3\text{-}90)$

From equation 3-52, it will be noted that

$$w = l_j{}^* Z_{Lj-Lk} l_k$$

Then
$$w' = \mathcal{R} l_j{}^* Z_{Lj-Lk} l_k \tag{3-91}$$

It is desirable to derive an alternative method of obtaining w' which would not require the measurement of all the Z_{Lj-Lk} impedances. For a system of 60 loads the number of Z_{Lj-Lk} impedances would be 3600. The calculation of w' for a system of 60 loads would thus be prohibitively lengthy. A much shorter method of calculating w' is discussed later.

The imaginary skew-symmetric part of Z_{mn} is given by

$$\frac{X_{m-n} - X_{n-m}}{2}$$

$$= \mathcal{I}\left[\frac{Z_{m-n} - Z_{n-m}}{2}\right]$$

$$= \mathcal{I}\left[\frac{Z_{Gm-Gn} - a_m - b_n + w - Z_{Gn-Gm} + a_n + b_m - w}{2}\right]$$

$$= \mathcal{I}\left[\frac{-(a_m - b_m) + (a_n - b_n)}{2}\right]$$

where $\mathcal{I}$ denotes imaginary part of quantity indicated. From equation 3-52

$$a_m = Z_{Gm-Lk} l_k$$
$$= Z_{Gm-Lk}(l_k' + j l_k'')$$
$$b_m = Z_{Lk-Gm} l_k{}^* = Z_{Gm-Lk} l_k{}^* = Z_{Gm-Lk}(l_k' - j l_k'')$$

Similarly,
$$a_n = Z_{Gn-Lk}(l_k' + j l_k'')$$
$$b_n = Z_{Gn-Lk}(l_k' - j l_k'')$$

Then
$$a_m - b_m = 2 Z_{Gm-Lk}(j l_k'')$$
$$\mathcal{I}(a_m - b_m) = 2 R_{Gm-Lk} l_k''$$

Similarly,
$$\mathcal{I}(a_n - b_n) = 2 R_{Gn-Lk} l_k''$$

Then
$$\frac{X_{m-n} - X_{n-m}}{2} = \left[\frac{-2 R_{Gm-Lk} l_k'' + 2 R_{Gn-Lk} l_k''}{2}\right]$$
$$= [-R_{Gm-Lk} l_k'' + R_{Gn-Lk} l_k'']$$
$$= [-f_m + f_n] \tag{3-92}$$

where
$$f_m = R_{Gm-Lk} l_k'' \tag{3-93}$$
$$f_n = R_{Gn-Lk} l_k''$$

DEVELOPMENT OF TRANSMISSION LOSS FORMULA

In terms of the system of equation 3-59 we have

$$\frac{X_{1-2} - X_{2-1}}{2} = -(R_{G1-L1}l_1'' + R_{G1-L2}l_2'')$$
$$+ (R_{G2-L1}l_1'' + R_{G2-L2}l_G'')$$

As previously noted, the skew-symmetry in the reactances results from terms involving the products of imaginary load currents and mutual resistances between generators and loads.

In summary we note that the transmission losses are given by

$$P_L = i_{dm}R_{m-n}i_{dn} + i_{qm}R_{m-n}i_{qn}$$
$$+ 2i_{dm}[f_m - f_n]i_{qn} \qquad (3\text{-}94)$$

where hereafter R_{m-n} will be understood to be the frame 3 symmetric resistance given by

$$R_{m-n} = R_{Gm-Gn} - d_n - d_m + w' \qquad (3\text{-}84)$$

where
$$d_m = R_{Gm-Lk}l_k' \qquad (3\text{-}90)$$

Also
$$f_m = R_{Gm-Lk}l_k'' \qquad (3\text{-}93)$$

3.9 TRANSFORMATION OF GENERATOR CURRENTS TO GENERATOR POWERS

Equation 3-94 expresses the transmission losses in terms of generator currents. The load dispatcher customarily works in terms of powers. Hence it is necessary for us to transform equation 3-94 to generator powers so that our expression for losses will be the most useful. The steps involved in proceeding from generator currents to generator powers are described as reference frames 4, 5, and 6 by Kron. In this section we shall proceed directly from reference frame 3 to reference frame 6.

Denote by θ_m the angle between the reference axis and the voltage of generator m. The reference axis is the common axis upon which all voltages and currents have been projected in all our previous work.

Let P_m = real power output of generator m

Q_m = reactive power output of generator m

V_m = absolute value of the voltage of generator m

From Figure 3.19, it will be seen that

$$i_{dm} = \frac{1}{V_m}[P_m \cos \theta_m + Q_m \sin \theta_m] \qquad (3\text{-}95)$$

$$i_{qm} = \frac{-1}{V_m}[-P_m \sin \theta_m + Q_m \cos \theta_m] \qquad (3\text{-}96)$$

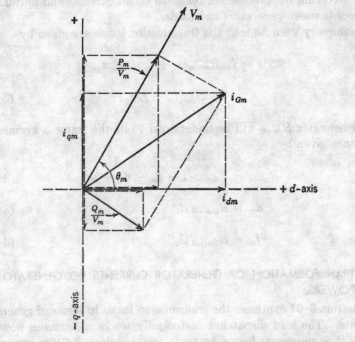

Figure 3.19. Vector diagram for trigonometric projections.

To eliminate Q_m as a variable, it is assumed that the ratio of Q_m/P_m will remain a constant value s_m.

Thus, equations 3-95 and 3-96 may be written

$$i_{dm} = \frac{1}{V_m}[\cos \theta_m + s_m \sin \theta_m]P_m \qquad (3\text{-}97)$$

$$i_{qm} = \frac{-1}{V_m}[-\sin \theta_m + s_m \cos \theta_m]P_m \qquad (3\text{-}98)$$

Substituting equations 3-97 and 3-98 into equation (3-94), we have

DEVELOPMENT OF TRANSMISSION LOSS FORMULA

$$P_L = P_m \left[\frac{1}{V_m} (\cos \theta_m + s_m \sin \theta_m) \right] R_{m-n}$$

$$\times \left[\frac{1}{V_n} (\cos \theta_n + s_n \sin \theta_n) \right] P_n$$

$$+ P_m \left[\frac{1}{V_m} (-\sin \theta_m + s_m \cos \theta_m) \right] R_{m-n}$$

$$\times \left[\frac{1}{V_n} (-\sin \theta_n + s_n \cos \theta_n) \right] P_n$$

$$- 2 P_m \left[\frac{1}{V_m} (\cos \theta_m + s_m \sin \theta_m) \right] [f_m - f_n]$$

$$\times \left[\frac{1}{V_n} (-\sin \theta_n + s_n \cos \theta_n) \right] P_n \tag{3-99}$$

$$= P_m \left[\frac{1}{V_m V_n} (\cos \theta_m \cos \theta_n + s_n \cos \theta_m \sin \theta_n \right.$$

$$\left. + s_m \sin \theta_m \cos \theta_n + s_m s_n \sin \theta_m \sin \theta_n) R_{m-n} \right] P_n$$

$$+ P_m \left[\frac{1}{V_m V_n} (\sin \theta_m \sin \theta_n - s_n \sin \theta_m \cos \theta_n \right.$$

$$\left. - s_m \cos \theta_m \sin \theta_n + s_m s_n \cos \theta_m \cos \theta_n) R_{m-n} \right] P_n$$

$$- 2 P_m \left[\frac{1}{V_m V_n} (-\cos \theta_m \sin \theta_n + s_n \cos \theta_m \cos \theta_n \right.$$

$$\left. - s_m \sin \theta_m \sin \theta_n + s_m s_n \sin \theta_m \cos \theta_n)(f_m - f_n) \right] P_n \tag{3-100}$$

Recall that $\cos \theta_{mn} = \cos (\theta_m - \theta_n)$

$$= \cos \theta_m \cos \theta_n + \sin \theta_m \sin \theta_n \tag{3-101}$$

$$\sin \theta_{mn} = \sin (\theta_m - \theta_n)$$

$$= \sin \theta_m \cos \theta_n - \sin \theta_n \cos \theta_m \tag{3-102}$$

Substituting equations 3–102 and 3–101 into 3–100, we have

$$P_L = P_m K_{mn} R_{m-n} P_n - 2 P_m F_{mn} P_n \tag{3-103}$$

where $\quad K_{mn} = \dfrac{1}{V_m V_n} [(1 + s_m s_n) \cos \theta_{mn} + (s_m - s_n) \sin \theta_{mn}] \tag{3-104}$

$$F_{mn} = \frac{1}{V_m V_n} (-\cos\theta_m \sin\theta_n + s_n \cos\theta_m \cos\theta_n - s_m \sin\theta_m \sin\theta_n$$
$$+ s_m s_n \sin\theta_m \cos\theta_n)(f_m - f_n) \quad (3\text{-}105)$$

The calculation for $-2P_m F_{mn} P_n$ may be simplified by recalling that only the symmetrical part of F_{mn} is required since $-2P_m F_{mn} P_n$ is a quadratic form, as discussed in Section 3.8.
Then

$$\frac{F_{mn} + F_{nm}}{2} = \frac{1}{2 V_m V_n} \left[\begin{array}{l} (-\cos\theta_m \sin\theta_n + s_n \cos\theta_m \cos\theta_n \\ - s_m \sin\theta_m \sin\theta_n + s_m s_n \sin\theta_m \cos\theta_n) \\ - (-\cos\theta_n \sin\theta_m + s_m \cos\theta_n \cos\theta_m \\ - s_n \sin\theta_n \sin\theta_m + s_n s_m \sin\theta_n \cos\theta_m) \end{array} \right] (f_m - f_n) \quad (3\text{-}106)$$

Substituting equations 3-101 and 3-102 into 3-106,

$$\frac{F_{mn} + F_{nm}}{2} = \frac{1}{2 V_m V_n} [(1 + s_m s_n) \sin\theta_{mn} + (s_n - s_m) \cos\theta_{mn}](f_m - f_n)$$
$$= \tfrac{1}{2} H_{mn}(f_m - f_n) \quad (3\text{-}107)$$

where $\quad H_{mn} = \dfrac{1}{V_m V_n} [(1 + s_m s_n) \sin\theta_{mn} + (s_n - s_m) \cos\theta_{mn}] \quad (3\text{-}108)$

Then equation 3-103 becomes

$$P_L = P_m K_{mn} R_{m-n} P_n - 2 P_m \frac{F_{mn} + F_{nm}}{2} P_n \quad (3\text{-}109)$$

$$= P_m K_{mn} R_{m-n} P_n - P_m H_{mn}(f_m - f_n) P_n \quad (3\text{-}110)$$

$$= P_m B_{mn} P_n$$

where $\quad B_{mn} = K_{mn} R_{m-n} - H_{mn}(f_m - f_n) \quad (3\text{-}111)$

If the term $H_{mn}(f_m - f_n)$ is neglected, the expression for B_{mn} reduces to

$$B_{mn} = K_{mn} R_{m-n} \quad (3\text{-}112)$$

The conditions for which $H_{mn}(f_m - f_n)$ may be negligibly small are discussed in Section 4.3 of Chapter 4.

The equivalent circuit corresponding to equation 3-110 is given in Figure 3.2. We now have impressed generator powers instead of generator currents. The B_{mn} represent an equivalent loss network through which the generator powers flow in supplying the over-all system load. Since $B_{mn} = B_{nm}$, the number of loss-formula coefficients to be calculated for a loss formula with n sources is $[n(n+1)]/2$.

DEVELOPMENT OF TRANSMISSION LOSS FORMULA 93

The relative magnitudes of the B_{mn} terms can be estimated from a physical knowledge of the transmission system. For example, those sources which are the greatest distance from the system load will have the largest value of the self B_{nn} terms. Sources close to each other usually have positive mutual terms, and sources on opposite ends of the system usually have negative mutuals. In any given row or column of coefficients the self term is always positive and generally the largest positive number in that row or column.

3.10 DETERMINATION OF w'

From equations 3–111 and 3–84 we may write

$$B_{mn} = K_{mn} R_{m-n} - H_{mn}(f_m - f_n)$$
$$= K_{mn}(R_{Gm-Gn} - d_n - d_m + w') - H_{mn}(f_m - f_n)$$
$$= K_{mn}(R_{Gm-Gn} - d_n - d_m) + K_{mn} w' - H_{mn}(f_m - f_n)$$
$$= A_{mn} + K_{mn} w' - H_{mn}(f_m - f_n) \qquad (3\text{–}113)$$

where
$$A_{mn} = K_{mn}(R_{Gm-Gn} - d_n - d_m) \qquad (3\text{–}114)$$

The losses from the loss formula may be equated to the base case losses obtained by summing the $i_k{}^2 R_k$ losses in all the lines where

i_k = scalar current in line k

R_k = resistance of line k

$$\sum i_k{}^2 R_k = P_m B_{mn} P_n$$
$$= P_m A_{mn} P_n + P_m K_{mn} w' P_n - P_m H_{mn}(f_m - f_n) P_n$$

Solving for w', we obtain

$$w' = \frac{\sum i_k{}^2 R_k - P_m A_{mn} P_n + P_m H_{mn}(f_m - f_n) P_n}{P_m K_{mn} P_n} \qquad (3\text{–}115)$$

Neglecting the effect of $H_{mn}(f_m - f_n)$, the expression for w' becomes

$$w' = \frac{\sum i_k{}^2 R_k - P_m A_{mn} P_n}{P_m K_{mn} P_n} \qquad (3\text{–}116)$$

Calculation of w' in this manner, as given by equation 3–115 or 3–116, eliminates the need for the determination of the self and mutual impedances of the loads (Z_{Lj-Lk}).

3.11 REPRESENTATION OF SOURCE-REACTIVE CHARACTERISTICS

In Section 3.9 Q_m was eliminated as a variable by assuming

$$Q_m = s_m P_m \qquad (3\text{-}117)$$

In many cases, particularly when s_m is small, this assumption results in satisfactory answers.

In this section we shall discuss other methods of considering the reactive power of the source. For example, consider the Q_m vs. P_m relationship assumed in Figure 3.20. It will be noted that $Q_1 = s_1 P_1$ would not

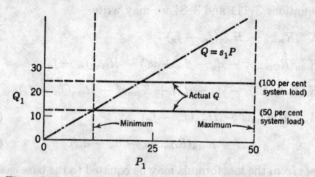

Figure 3.20. Reactive power plotted as function of real power.

be a very accurate representation of the variation of Q_1 with P_1. In this particular case Q_1 is independent of P_1, but it is determined by the total system-load level. Thus it would be appropriate to include the plant reactive as part of the load of that bus. As the system load increases, Q_1 increases. Similarly, when the system load decreases, Q_1 decreases. Denote by Q_{Lm} that part of reactive power of plant m which is included as part of the load at that bus.

Then
$$Q_m = Q_{Lm} + s_m P_m \qquad (3\text{-}118)$$

For Figure 3.20 we have

$$Q_1 = Q_{L1} + s_1 P_1 = Q_{L1} \qquad \text{since } s_1 = 0$$

Another case of interest is given in Figure 3.21. It is seen that Q_2 is a function of the total system load as well as P_2. In this case we have at 100 per cent system load

$$Q_2 = Q_{L2} + s_2 P_2$$
$$= 100 - 0.25 P_2$$

At 50 per cent system load

$$Q_2 = 50 - 0.25P_2$$

By assuming Q_m as a function of system load and plant output, considerable flexibility is available in representing the source reactive characteristics appropriately.

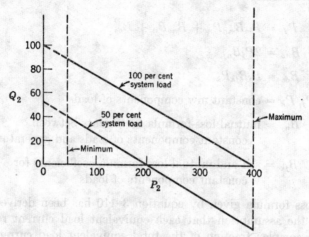

Figure 3.21. Reactive power plotted as function of real power.

The determination of the reactive characteristics discussed requires a plot of field data or data from load-flow studies. If no data is available, it is suggested that the value of s_m be approximated by

$$s_m = -\frac{R}{X} = \frac{\Delta Q_m}{\Delta P_m} \qquad (3\text{-}119)$$

where R/X pertains to the ratio of resistance to reactance of the transmission system. This value of s_m approximately indicates the change in mvar loading as a function of mw loading for the condition of maintaining constant scalar voltages at both ends of a transmission line.

If Q_m is a variable such that it cannot be appropriately represented as a function of system load and plant output, it may be retained as an independent variable in the loss formula. In this case the work of Section 3.9 would be carried through by substituting equations 3-95 and 3-96 into 3-94.

3.12 REPRESENTATION OF LOADS

On occasion, loads such as aluminum or paper mill loads are encountered which do not vary over the daily load cycle in the same manner as the rest of the system loads. However, these loads may be included properly in the loss formula as negative sources.

Also, it is sometimes desirable to divide the loads at the various buses into a component which varies with the total load and a component which remains constant. The constant components are treated as negative generators in the loss formula. The loss formula then takes this form:

$$P_L = P_m B_{mn} P_n + B_{no} P_n + B_{oo} \tag{3-120}$$

where $B_{no} = 2P_j B_{nj}$

$B_{oo} = P_j B_{jk} P_k$

P_j, P_k = constant mw components of loads

B_{nj} = mutual loss-formula coefficients between constant components of loads and generators

B_{jk} = self and mutual loss-formula coefficients for constant components of loads

The loss formula given by equation 3–110 has been derived on the basis of the assumption that each equivalent load current remains a constant complex fraction of the total equivalent load current. This assumption has proved adequate for many system studies. Equation 3–110 may be extended [5] to the case in which each individual load current is assumed to be a linear complex function of the total load current. Thus, the assumption that

$$i_{Lj} = l_j i_L \tag{3-35}$$

is to be replaced by the assumption that

$$i_{Lj} = l_j i_L + i_j^o \tag{3-121}$$

where l_j denotes the complex rate of change of load current j with the total load current and i_j^o denotes the value of load current j when the sum of all load currents equals zero.

By following a procedure similar to that outlined in Sections 3.6 to 3.9, inclusive, we again obtain a formula of the form given by equation 3–120. The quantities B_{mn}, B_{no}, and B_{oo} are then given as

$$B_{mn} = A_{mn} + K_{mn} w' - H_{mn}(f_m - f_n) \tag{3-113}$$

where $A_{mn} = K_{mn}(R_{Gm-Gn} - d_n - d_m) \tag{3-114}$

$d_m = R_{Gm-Lk} l_k' \tag{3-90}$

$f_m = R_{Gm-Lk} l_k'' \tag{3-93}$

$w' = \mathcal{R} l_j^* Z_{Lj-Lk} l_k \tag{3-91}$

$$K_{mn} = \frac{1}{V_m V_n}[(1 + s_m s_n)\cos\theta_{mn} + (s_m - s_n)\sin\theta_{mn}] \quad (3\text{-}104)$$

$$H_{mn} = \frac{1}{V_m V_n}[(1 + s_m s_n)\sin\theta_{mn} + (s_n - s_n)\cos\theta_{mn}] \quad (3\text{-}108)$$

$$B_{no} = 2P_{Lk}{}^{o}[(R_{Gn-Lk} - l_j' R_{Lj-Lk})K_{nk} + (l_j'' R_{Lj-Lk})H_{nk}] \quad (3\text{-}122)$$

where $\quad P_{Lk}{}^{o} =$ real power value of load k when $i_L = 0 \quad (3\text{-}123)$

$$K_{nk} = \frac{1}{V_n V_{Lk}}[(1 + s_n s_{Lk})\cos\theta_{n-Lk} + (s_n - s_{Lk})\sin\theta_{n-Lk}] \quad (3\text{-}124)$$

$$H_{nk} = \frac{1}{V_n V_{Lk}}[(1 + s_n s_{Lk})\sin\theta_{n-Lk} + (s_{Lk} - s_n)\cos\theta_{n-Lk}] \quad (3\text{-}125)$$

Also

$$B_{oo} = \mathcal{R}(i_j{}^{o*} Z_{Lj-Lk} i_k{}^{o}) \quad (3\text{-}126)$$

It will be noted that the expression for B_{mn} is identical to that obtained previously, with the exception that l_j is now defined by equation (3-121). The determination of B_{no} and B_{oo} requires the measurement of all the self and mutual impedances between loads. Thus for a system with 100 loads it would be required to determine 5050 self- and mutual-load impedances. The determination of this many load impedances is generally prohibitively lengthy. A more practical method [6] of calculating the B_{no} and B_{oo} terms which does not require the determination of these self- and mutual-load impedances is described in Chapter 4. This involves the use of circuit theory together with a least-squares solution for w', B_{no}, and B_{oo}.

3.13 REACTIVE LOSSES

A formula for reactive losses may be derived by taking the imaginary part of $I_3^* E_3$ and proceeding according to Sections 3.8 and 3.9. The reactive losses are given by

reactive losses

$$= i_{dm} X_{m-n} i_{dn} + i_{qm} X_{m-n} i_{qn}$$

$$+ 2i_{dm}\left(\frac{R_{m-n} - R_{n-m}}{2}\right) i_{qn}$$

$$= P_n B_{mn}' P_n \quad (3\text{-}127)$$

where

$$B_{mn}' = K_{mn}[X_{Gm-Gn} - X_{Gm-Lk}l_k' - l_j'X_{Lj-Gn} - l_j'X_{Lj-Lk}l_k]$$
$$+ H_{mn}[X_{Gm-Lk}l_k'' - l_j''X_{Lj-Gn}] \qquad (3\text{--}128)$$

3.14 CONSIDERATION OF OFF-NOMINAL TURN RATIOS [7,8]

When the turn-ratios of transformers differ from the ratios of nominal values assigned to the different levels of transmission the representation of a transmission system requires the use of auto-transformers. These auto-transformers provide a path to ground which must be considered in the derivation of the loss-formula coefficients.

A detailed description of the methods of calculating auto-transformer tap settings is given in Chapter 5 of Reference 9. From this reference it

Figure 3.22. Schematic representation of network analyzer auto-transformer (variable end toward high-voltage bus).

is seen that if the variable end is considered to be towards the high-voltage bus, as indicated in Figure 3.22, then,

$$\text{setting} = \frac{V_{\text{base }L}}{V_{\text{base }H}} \times \frac{V_{\text{trans }H}}{V_{\text{trans }L}} - 1$$

where $V_{\text{base }H}$ = system-base voltage on the high-voltage side of the transformer

$V_{\text{base }L}$ = system-base voltage on the low-voltage side of the transformer

$V_{\text{trans }H}$ = transformer rated high voltage for the particular tap being considered

$V_{\text{trans }L}$ = transformer rated low voltage for the particular tap being considered

The arrow is used to indicate the variable end.

For example, assume the following situation:

1. Transformer tap position corresponds to 70.725/13.8 kv.
2. High-voltage base = 66 kv.
3. Low-voltage base = 13 kv.

Then the setting is given by the above equation as

$$\text{setting} = \frac{13}{66} \times \frac{70.725}{13.8} - 1$$
$$= 1.01 - 1 = 0.01$$

DEVELOPMENT OF TRANSMISSION LOSS FORMULA

This result is shown in Figure 3.23.

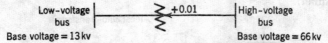

Figure 3.23. Setting for example given in Section 3.14.

If the variable end is considered to be toward the low-voltage bus, as indicated in Figure 3.24, the expression for the setting is then given by

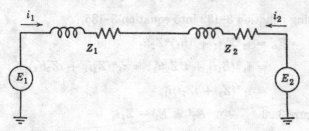

Figure 3.24. Schematic representation with variable end toward low-voltage bus.

$$\text{setting} = \frac{V_{\text{base } H}}{V_{\text{base } L}} \times \frac{V_{\text{trans } L}}{V_{\text{trans } H}} - 1$$

Consider the circuit given in Figure 3.25. For this circuit we may

Figure 3.25. Simple circuit without auto-transformer.

write by inspection that

$$E_1 - E_2 = Z_{11} i_1 \tag{3-129}$$

and
$$P_L = i_1^*(E_1 - E_2) = i_1^* Z_{11} i_1 \tag{3-130}$$

where
$$Z_{11} = Z_1 + Z_2 \tag{3-131}$$

Next consider the simple circuit of Figure 3.26 in which we have inserted an ideal auto-transformer. We shall proceed to discover the manner in which equations 3–129, 3–130, and 3–131 are modified by this auto-transformer. For Figure 3.26 we note that

$$i_2 = -t i_1 \quad \text{where} \quad t = 1 + \text{setting} \tag{3-132}$$

$$i_t = -(i_1 + i_2) = -(i_1 - t i_1)$$

$$= -i_1(1 - t) \tag{3-133}$$

$$E_2' = \frac{E_1'}{t} \tag{3-134}$$

The magnetizing current is assumed to be zero and t is assumed to be a real number.

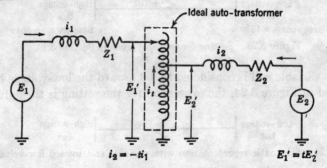

Figure 3.26. Simple circuit with ideal auto-transformer.

The losses in the circuit of Figure 3.26 are given by

$$P_L = i_1{}^*Z_1 i_1 + i_2{}^*Z_2 i_2 \qquad (3\text{–}135)$$

Substituting equation 3–132 into equation 3–135,

$$P_L = i_1{}^*Z_1 i_1 + (ti_1)^*Z_2 ti_1$$
$$= i_1{}^*(Z_1 i_1 + t^*Z_2 ti_1) = i_1{}^*(Z_1 i_1 + tZ_2 ti_1)$$
$$= i_1{}^*(Z_1 + tZ_2 t)i_1 \qquad (3\text{–}136)$$

From Figure 3.26
$$E_1' = E_1 - Z_1 i_1 \qquad (3\text{–}137)$$
$$E_2' = E_2 - Z_2 i_2 \qquad (3\text{–}138)$$

Substituting equations 3–132 and 3–134 into equation 3–138, we obtain

$$\frac{E_1'}{t} = E_2 + Z_2 ti_1 \qquad (3\text{–}139)$$

Multiplying equation 3–139 by t and subtracting from equation 3–137

$$E_1' - E_1' = E_1 - tE_2 - (Z_1 i_1 + tZ_2 ti_1) = 0$$

and
$$E_1 - tE_2 = Z_1 i_1 + tZ_2 ti_1 = (Z_1 + tZ_2 t)i_1 \qquad (3\text{–}140)$$

Thus, from equations 3–136 and 3–140

$$P_L = i_1{}^*(E_1 - tE_2) \qquad (3\text{–}141)$$

If point 2 is grounded and 1 is energized, it will be seen that

$$\frac{E_1}{i_1} = Z_{11} = Z_1 + t^2 Z_2$$

and
$$t = -i_2/i_1$$

DEVELOPMENT OF TRANSMISSION LOSS FORMULA

In summary we have

$$E_1 - tE_2 = Z_{11}i_1 \qquad (3\text{-}142)$$

$$P_L = i_1{}^*(E_1 - tE_2) = i_1{}^*Z_{11}i_1 \qquad (3\text{-}143)$$

and
$$Z_{11} = Z_1 + t^2 Z_2 \qquad (3\text{-}144)$$

$$i_t = -i_1(1 - t) \qquad (3\text{-}133)$$

By comparison of equations 3-142, 3-143, and 3-144 with equations 3-129, 3-130, and 3-131, we may note the manner in which the presence of an auto-transformer has modified our equations. The impedance Z_{11} may be determined by the same measurement as before. The expression for the voltage of interest is now $E_1 - tE_2$ instead of $E_1 - E_2$.

In a similar manner it can be shown that the performance of the transmission system in reference frame 2 may be closely approximated by

$$
\begin{array}{c c}
 & \begin{array}{cc} Gn & L \end{array} \\
\begin{array}{c} Gm \\ L \end{array} \begin{array}{|c|} \hline E_{Gm} - t_m E_R \\ \hline E_L - t_L E_R \\ \hline \end{array}
=
\begin{array}{c} Gm \\ L \end{array} \begin{array}{|c|c|} \hline Z_{Gm\text{-}Gn} & a_m \\ \hline b_n & w \\ \hline \end{array}
\begin{array}{c} Gn \\ L \end{array} \begin{array}{|c|} \hline i_{Gn} \\ \hline i_L \\ \hline \end{array}
\end{array}
\qquad (3\text{-}145)
$$

where t_m and t_L are turn-ratio constants which are considered to be real numbers. The other quantities are defined and measured as previously discussed. The circuit corresponding to equation 3-145 is given in Figure 3.27. The use of this equation has been found sufficiently accurate

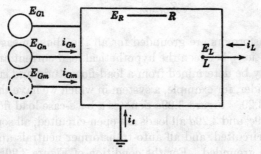

Figure 3.27. Reference frame 2 with auto-transformers.

for considering transmission systems as usually designed. A more rigorous analysis is given in Reference 7.

The term t_n represents the equivalent turn ratio of the hypothetical auto-transformer that exists between generator G_n and the reference point R. Similarly, t_L represents the equivalent turn ratio of the hypothetical auto-transformer that exists between the hypothetical load and the reference point. The current i_t is the total turn-ratio current that exists in this path to ground.

The turn ratio t_n between generator G_n and reference point R may be determined when the frame 1 measurements are taken. Referring to Figure 3.28,

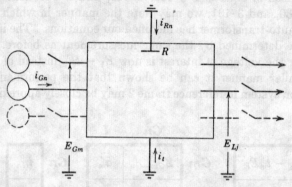

Figure 3.28. Measurement of impedances and equivalent turn ratio.

$$t_n = \frac{-i_{Rn}}{i_{Gn}} \qquad (3\text{-}146)$$

As before, $$Z_{Gm-Gn} = \frac{E_{Gm}}{i_{Gn}} \qquad (3\text{-}8)$$

$$Z_{Lj-Gn} = \frac{E_{Lj}}{i_{Gn}} \qquad (3\text{-}9)$$

The auto-transformers are grounded for all the above measurements.

The turn ratio t_L between the hypothetical load current and the reference point may be determined from a load-flow study and superposition theory. Consider, for example, a system in which we have three sources, as in Figure 3.29. Figure 3.29a denotes a base-case load flow. In Figures 3.29b, 3.29c, and 3.29d all loads are open-circuited, all sources except one are open-circuited, and all auto-transformer neutrals and the reference point are grounded. For the condition of Figure 3.29b impress on generator 1 a current i_{G1} corresponding to the value existing in the base case of Figure 3.29a. The current i_{R1} in this case is given by

$$i_{R1} = -t_1 i_{G1}$$

DEVELOPMENT OF TRANSMISSION LOSS FORMULA

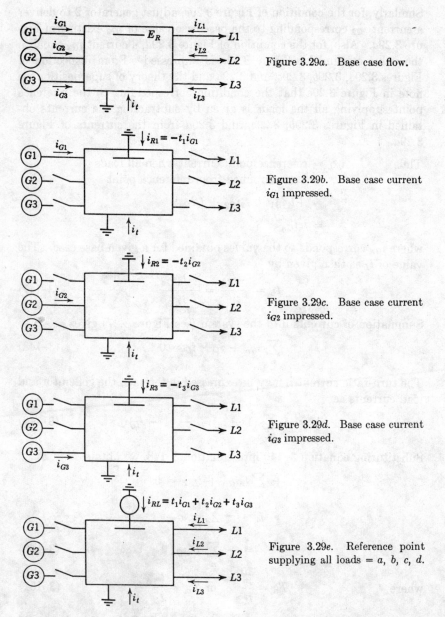

Figure 3.29a. Base case flow.

Figure 3.29b. Base case current i_{G1} impressed.

Figure 3.29c. Base case current i_{G2} impressed.

Figure 3.29d. Base case current i_{G3} impressed.

Figure 3.29e. Reference point supplying all loads = a, b, c, d.

Figure 3.29. Determination of turn-ratio t_L by means of superposition.

Similarly, for the condition of Figure 3.29c, adjust generator 2 to deliver a current i_{G2} corresponding to the base-case value of i_{G2} existing in Figure 3.29a. Also, for the condition of Figure 3.29d, a current i_{G3} equal to the value existing for Figure 3.29a is impressed. From inspection of Figures 3.29a, 3.29b, 3.29c, and 3.29d and the theory of superposition we note in Figure 3.29e that the condition corresponding to the reference point supplying all the loads is given by subtracting the currents obtained in Figures 3.29b, 3.29c, and 3.29d from the currents of Figure 3.29a.

Thus $\quad i_{RL} = $ reference point current when all loads are supplied from reference point

$$= +t_1 i_{G1} + t_2 i_{G2} + t_3 i_{G3}$$

$$= t_n i_{Gn}$$

where i_{Gn} corresponds to the values obtained for a given base case. The value of t_L is then given by

$$t_L = \frac{-i_{RL}}{i_L} = \frac{-t_n i_{Gn}}{i_L} \tag{3-147}$$

Summation of currents into the system (see Figure 3.27) gives

$$\sum_n i_{Gn} + i_t + i_L = 0 \tag{3-148}$$

The turn-ratio current i_t may be expressed in terms of the generator and load currents as

$$i_t = -\sum_n (1 - t_n) i_{Gn} - (1 - t_L) i_L \tag{3-149}$$

Substituting equation 3-149 into equation 3-148, we obtain

$$+ \sum_n t_n i_{Gn} + t_L i_L = 0$$

$$i_L = -\sum_n \frac{t_n}{t_L} i_{Gn}$$

$$= -\sum_n T_n i_{Gn} \tag{3-150}$$

where $\quad T_n = \dfrac{t_n}{t_L} \quad$ or $\quad T_m = \dfrac{t_m}{t_L} \tag{3-151}$

It will be recalled that for cases previously considered in this chapter in which no auto-transformers are present $T_n = 1$ for all values of n.

For a system with three sources we may write the transformation from reference frame 2 to reference frame 3 as

DEVELOPMENT OF TRANSMISSION LOSS FORMULA

$$C_3{}^2 = \begin{array}{c|ccc} & G1 & G2 & G3 \\ \hline G1 & 1 & & \\ G2 & & 1 & \\ G3 & & & 1 \\ \hline L & -T_1 & -T_2 & -T_3 \end{array} \qquad (3\text{-}152)$$

Recalling that the reference frame 2 impedance is given by

$$\begin{array}{c|cccc} & G1 & G2 & G3 & L \\ \hline G1 & Z_{G1-G1} & Z_{G1-G2} & Z_{G1-G3} & a_1 \\ G2 & Z_{G2-G1} & Z_{G2-G2} & Z_{G2-G3} & a_2 \\ G3 & Z_{G3-G1} & Z_{G3-G2} & Z_{G3-G3} & a_3 \\ \hline L & b_1 & b_2 & b_3 & w \end{array} \qquad (3\text{-}42)$$

we may obtain the reference frame 3 impedances by the following steps:

$$Z_{\text{old}}C = \begin{array}{c|cccc} & G1 & G2 & G3 & L \\ \hline G1 & Z_{G1-G1} & Z_{G1-G2} & Z_{G1-G3} & a_1 \\ G2 & Z_{G2-G1} & Z_{G2-G2} & Z_{G2-G3} & a_2 \\ G3 & Z_{G3-G1} & Z_{G3-G2} & Z_{G3-G3} & a_3 \\ \hline L & b_1 & b_2 & b_3 & w \end{array} \quad \begin{array}{c|ccc} & G1 & G2 & G3 \\ \hline G1 & 1 & & \\ G2 & & 1 & \\ G3 & & & 1 \\ \hline L & -T_1 & -T_2 & -T_3 \end{array}$$

$$= \begin{array}{c|ccc} & G1 & G2 & G3 \\ \hline G1 & Z_{G1-G1} - a_1 T_1 & Z_{G1-G2} - a_1 T_2 & Z_{G1-G3} - a_1 T_3 \\ G2 & Z_{G2-G1} - a_2 T_1 & Z_{G2-G2} - a_2 T_2 & Z_{G2-G3} - a_2 T_3 \\ G3 & Z_{G3-G1} - a_3 T_1 & Z_{G3-G2} - a_3 T_2 & Z_{G3-G3} - a_3 T_3 \\ \hline L & b_1 - wT_1 & b_2 - wT_2 & b_3 - wT_3 \end{array}$$

and

$C_t {}^* Z_{\text{old}} C =$

	G1	G2	G3	L
G1	1			$-T_1$
G2		1		$-T_2$
G3			1	$-T_3$

	G1	G2	G3
G1	$Z_{G1-G1} - a_1 T_1$	$Z_{G1-G2} - a_1 T_2$	$Z_{G1-G3} - a_1 T_3$
G2	$Z_{G2-G1} - a_2 T_1$	$Z_{G2-G2} - a_2 T_2$	$Z_{G2-G3} - a_2 T_3$
G3	$Z_{G3-G1} - a_3 T_1$	$Z_{G3-G2} - a_3 T_2$	$Z_{G3-G3} - a_3 T_3$
L	$b_1 - w T_1$	$b_2 - w T_2$	$b_3 - w T_3$

$$= \begin{array}{c|ccc} & G1 & G2 & G3 \\ \hline G1 & \begin{array}{c} Z_{G1-G1} - a_1 T_1 \\ - T_1 b_1 \\ + T_1 w T_1 \end{array} & \begin{array}{c} Z_{G1-G2} - a_1 T_2 \\ - T_1 b_2 \\ + T_1 w T_2 \end{array} & \begin{array}{c} Z_{G1-G3} - a_1 T_3 \\ - T_1 b_3 \\ + T_1 w T_3 \end{array} \\ \hline G2 & \begin{array}{c} Z_{G2-G1} - a_2 T_1 \\ - T_2 b_1 \\ + T_2 w T_1 \end{array} & \begin{array}{c} Z_{G2-G2} - a_2 T_2 \\ - T_2 b_2 \\ + T_2 w T_2 \end{array} & \begin{array}{c} Z_{G2-G3} - a_2 T_3 \\ - T_2 b_3 \\ + T_2 w T_3 \end{array} \\ \hline G3 & \begin{array}{c} Z_{G3-G1} - a_3 T_1 \\ - T_3 b_1 \\ + T_3 w T_1 \end{array} & \begin{array}{c} Z_{G3-G2} - a_3 T_2 \\ - T_3 b_2 \\ + T_3 w T_2 \end{array} & \begin{array}{c} Z_{G3-G3} - a_3 T_3 \\ - T_3 b_3 \\ + T_3 w T_3 \end{array} \end{array} \quad (3\text{-}153)$$

The reference frame 3 voltages are given by

$C_t {}^* C_{\text{old}} =$

	G1	G2	G3	L
G1	1			$-T_1$
G2		1		$-T_2$
G3			1	$-T_3$

G1	$E_{G1} - t_1 E_R$
G2	$E_{G2} - t_2 E_R$
G3	$E_{G3} - t_3 E_R$
L	$E_L - t_L E_R$

G1	$E_{G1} - t_1 E_R - T_1 E_L + T_1 t_L E_R$
G2	$E_{G2} - t_2 E_R - T_2 E_L + T_2 t_L E_R$
G3	$E_3 - t_3 E_R - T_3 E_L + T_3 t_L E_R$

(3-154)

DEVELOPMENT OF TRANSMISSION LOSS FORMULA

From equation 3-151

$$T_1 = \frac{t_1}{t_L}$$

$$T_2 = \frac{t_2}{t_L}$$

$$T_3 = \frac{t_3}{t_L}$$

Then

$$E_{G1} - t_1 E_R - T_1 E_L + T_1 t_L E_R = E_{G1} - T_1 E_L - (T_1 t_L - t_1) E_R$$
$$= E_{G1} - T_1 E_L - (t_1 - t_1) E_R$$
$$= E_{G1} - T_1 E_L$$

Thus our reference frame 3 voltages become

$$\begin{array}{|c|c|} \hline G1 & E_{G1} - T_1 E_L \\ \hline G2 & E_{G2} - T_2 E_L \\ \hline G3 & E_{G3} - T_3 E_L \\ \hline \end{array} \qquad (3\text{-}155)$$

For the general case the transformation from reference frame 2 to reference frame 3 is then given by

$$C_3{}^2 = \begin{array}{c|c} & Gn \\ \hline Gn & 1 \\ \hline L & -T_n \\ \end{array} \qquad (3\text{-}156)$$

The corresponding reference frame 3 quantities are obtained by the transformations indicated in equations 3-22, 3-23, and 3-24 applied to equations 3-145 and 3-156.
Thus

$$Z_{\text{old}} C = \begin{array}{|c|c|} \hline Z_{Gm-Gn} & a_m \\ \hline b_n & w \\ \hline \end{array} \begin{array}{|c|} \hline 1 \\ \hline -T_n \\ \hline \end{array} = \begin{array}{|c|} \hline Z_{Gm-Gn} - a_m T_n \\ \hline b_n - w T_n \\ \hline \end{array}$$

ECONOMIC OPERATION OF POWER SYSTEMS

and

$$C_t{}^*(Z_{\text{old}}C) = \begin{bmatrix} 1 & -T_m \end{bmatrix} \begin{bmatrix} Z_{Gm-Gn} - a_m T_n \\ b_n - wT_n \end{bmatrix}$$

$$= \begin{bmatrix} Z_{Gm-Gn} - a_m T_n - T_m b_n + T_m w T_n \end{bmatrix}$$
$$= Z_{m-n} \tag{3-157}$$

Also,

$$C_t{}^* E_{\text{old}} = \begin{bmatrix} 1 & -T_m \end{bmatrix} \begin{bmatrix} E_{Gm} - t_m E_R \\ E_L - t_L E_R \end{bmatrix}$$

$$= \begin{bmatrix} E_{Gm} - t_m E_R - T_m E_L + T_m t_L E_R \end{bmatrix} \tag{3-158}$$

From equation 3-151

$$T_m = \frac{t_m}{t_L}$$

Substituting equation 3-151 into 3-158,

$$E_{Gm} - T_m E_L + (T_m t_L - t_m) E_R = E_{Gm} - T_m E_L + (t_m - t_m) E_R$$
$$= E_{Gm} - T_m E_L \tag{3-159}$$

The reference frame 3 equations are then given by

$$\begin{bmatrix} E_{Gm} - T_m E_L \end{bmatrix} = \begin{bmatrix} Z_{Gm-Gn} - a_m T_n - T_m b_n + T_m w T_n \end{bmatrix} \begin{bmatrix} i_{Gn} \end{bmatrix}$$

$$= \begin{bmatrix} Z_{m-n} \end{bmatrix} \begin{bmatrix} i_{Gn} \end{bmatrix} \tag{3-160}$$

DEVELOPMENT OF TRANSMISSION LOSS FORMULA

If the losses in this equivalent circuit are evaluated as indicated in Section 3.8, we obtain

$$P_L = i_{dm}R_{m-n}i_{dn} + i_{qm}R_{m-n}i_{qn} - 2i_{dm}\left[\frac{X_{m-n} - X_{n-m}}{2}\right]i_{qn} \quad (3\text{-}82)$$

where
$$\begin{aligned}R_{m-n} &= \Re\tfrac{1}{2}[Z_{m-n} + Z_{n-m}] \\ &= \Re\tfrac{1}{2}[Z_{Gm-Gn} - a_m T_n - T_m b_n + T_m w T_n \\ &\quad + Z_{Gn-Gm} - a_n T_m - T_n b_m + T_n w T_m] \\ &= R_{Gm-Gn} - D_m - D_n + W_{mn}' \quad (3\text{-}161)\end{aligned}$$

where
$$D_m = \Re\frac{a_m T_n - T_n b_m}{2} = T_n R_{Gm-Lk}l_k' = T_n d_m \quad (3\text{-}162)$$

$$W_{mn}' = \Re\tfrac{1}{2}(T_m w T_n + T_n w T_m) = T_m w' T_n \quad (3\text{-}163)$$

Similarly,

$$\begin{aligned}\frac{X_{m-n} - X_{n-m}}{2} &= \mathscr{g}\left(\frac{Z_{m-n} - Z_{n-m}}{2}\right) \quad (3\text{-}164) \\ &= \mathscr{g}\tfrac{1}{2}[Z_{Gm-Gn} - a_m T_n - T_m b_n + T_m w T_n \\ &\quad - Z_{Gm-Gn} + a_n T_m - T_n b_m - T_n w T_m] \\ &= \mathscr{g}\tfrac{1}{2}[-(a_m T_n - T_n b_m) + (a_n T_m - T_m b_n) \\ &\quad + (T_m w T_n - T_n w T_m)] \\ &= (F_m - F_n) = T_n f_m - T_m f_n \quad (3\text{-}165)\end{aligned}$$

The generator currents may be projected to generator powers as indicated in Section 3.9. Summarizing, we have the following results when auto-transformers are involved:

$$B_{mn} = K_{mn}R_{m-n} - H_{mn}(T_n f_m - T_m f_n) \quad (3\text{-}166)$$

where
$$R_{m-n} = R_{Gm-Gn} - T_n d_m - T_m d_n + T_m w' T_n \quad (3\text{-}167)$$

$$T_m = \Re\frac{t_m}{t_L} \quad (3\text{-}168)$$

The quantities K_{mn} and H_{mn} are given by equations 3-104 and 3-108, respectively.

3.15 SUMMARY

The transmission losses may be closely approximated by means of a transmission loss formula of the form

$$P_L = P_m B_{mn} P_n$$

when B_{mn} = loss formula coefficients

P_m = source powers

The loss-formula coefficients may be considered as an equivalent transmission loss circuit from each generating source to the hypothetical load as shown in Figure 3.2.

The assumptions involved in deriving a loss formula of this form are

1. The equivalent load current at any bus remains a constant complex fraction of the total equivalent load current. Nonconforming loads may be treated as negative sources in the formula or in special cases may be handled by a loss formula including linear terms and a constant term in addition to the quadratic terms $(P_m B_{mn} P_n)$.
2. The generator-bus voltage magnitudes are assumed to remain constant.
3. The generator-bus angles are assumed to remain constant.
4. The source reactive power may be approximated by the sum of a component which varies with the system load and a component which varies with the source output.

APPENDIX 3

It is to be demonstrated that if

$$\boldsymbol{i}_{\text{old}} = \boldsymbol{C}\boldsymbol{i}_{\text{new}} \qquad (3\text{-}22)$$

and if the power is to remain invariant, the new set of voltages is given by

$$\boldsymbol{e}_{\text{new}} = \boldsymbol{C}_t{}^* \boldsymbol{e}_{\text{old}} \qquad (3\text{-}23)$$

and the new set of impedances is given by

$$\boldsymbol{Z}_{\text{new}} = \boldsymbol{C}_t{}^* \boldsymbol{Z}_{\text{old}} \boldsymbol{C} \qquad (3\text{-}24)$$

We have

$$\boldsymbol{e}_{\text{old}} = \boldsymbol{Z}_{\text{old}} \boldsymbol{i}_{\text{old}} \qquad (3\text{-}169)$$

$$\boldsymbol{P}_{\text{old}} = \boldsymbol{e}_{\text{old}} \boldsymbol{i}_{\text{old}}{}^* \qquad (3\text{-}170)$$

$$\boldsymbol{e}_{\text{new}} = \boldsymbol{Z}_{\text{new}} \boldsymbol{i}_{\text{new}} \qquad (3\text{-}171)$$

$$\boldsymbol{P}_{\text{new}} = \boldsymbol{e}_{\text{new}} \boldsymbol{i}_{\text{new}}{}^* \qquad (3\text{-}172)$$

However, it is required that

$$P_{\text{old}} = P_{\text{new}}$$

Hence,
$$e_{\text{old}} i_{\text{old}}^* = e_{\text{new}} i_{\text{new}}^*$$

Since
$$i_{\text{old}} = C i_{\text{new}}$$

we have
$$e_{\text{new}} i_{\text{new}}^* = e_{\text{old}} C^* i^*_{\text{new}}$$

$$e_{\text{new}} = e_{\text{old}} C^* \quad (3\text{-}23)$$

$$= C_t^* e_{\text{old}}$$

which demonstrates relation 3–23.

We shall now consider the derivation of equation 3–24.

We start with
$$e_{\text{old}} = Z_{\text{old}} i_{\text{old}}$$

From equation 3–22 we may write

$$e_{\text{old}} = Z_{\text{old}} C i_{\text{new}} \quad (3\text{-}173)$$

Multiplying both sides of 3–173 by C_t^*, we obtain

$$C_t^* e_{\text{old}} = C_t^* Z_{\text{old}} C i_{\text{new}} \quad (3\text{-}174)$$

From equations 3–23 and 3–174

$$e_{\text{new}} = C_t^* e_{\text{old}} = C_t^* Z_{\text{old}} C i_{\text{new}} \quad (3\text{-}175)$$

By inspection of equations 3–171 and 3–175,

$$Z_{\text{new}} = C_t^* Z_{\text{old}} C \quad (3\text{-}24)$$

References

1. *Tensorial Analysis of Integrated Transmission Systems—Part I: The Six Basic Reference Frames*, G. Kron. *AIEE Trans.*, Vol. 70, Part I, 1951, pp. 1239–1248.
2. *Analysis of Total and Incremental Losses in Transmission Systems*, L. K. Kirchmayer, G. W. Stagg. *AIEE Trans.*, Vol. 70, Part I, 1951, pp. 1197–1205.
3. *Tensor Analysis of Networks*, G. Kron. John Wiley and Sons, New York, 1939.
4. *Matrix Analysis of Electrical Networks*, P. LeCorbeiller. Harvard University Press, Cambridge, Massachusetts, John Wiley and Sons, New York, 1950.
5. *A New Method of Determining Constants for the General Transmission Loss Equation*, E. D. Early, R. E. Watson. *AIEE Trans.*, Vol. 74, Part III, 1955, pp. 1417–1421.
6. *Discussion of Reference 5*, A. F. Glimn, L. K. Kirchmayer, R. D. Wood. *AIEE Trans.*, Vol. 74, Part III, 1955, pp. 1421.
7. *Tensorial Analysis of Integrated Transmission Systems—Part II: Off-Nominal Turn-Ratios*, G. Kron. *AIEE Trans.*, Vol. 71, Part III, 1952, pp. 505–12.
8. *Transmission Losses and Economic Loading of Power Systems*, L. K. Kirchmayer, G. H. McDaniel. *General Electric Review*, Schenectady, New York, October 1951.

9. *General Electric Network Analyzers*, L. W. Robbins, F. S. Rothe. General Electric Company, Schenectady, New York, 1950.

Problems

Problem 3.1

Figure 4.2 is a simplified one-line impedance diagram of the Indiana Division of the American Gas and Electric Service Corporation.

Choose the Muncie bus as the reference bus and calculate the open-circuit self and mutual impedances as indicated in the following:

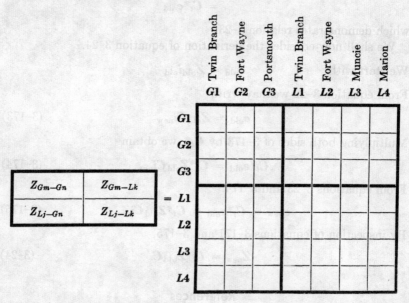

Problem 3.2

Perform the following matrix operations:

1. DD^{-1}

$$D = \begin{array}{c|c|c} & 1 & 2 \\ \hline 1 & 1 & 2 \\ \hline 2 & 3 & 4 \end{array}$$

$$D^{-1} = \begin{array}{c|c|c} & 1 & 2 \\ \hline 1 & -2 & +1 \\ \hline 2 & +1.5 & -.5 \end{array}$$

DEVELOPMENT OF TRANSMISSION LOSS FORMULA

2. FG
3. GF

where

$$F = \begin{array}{c|cc} & 1 & 2 \\ \hline 1 & 1 & 2 \\ 2 & 3 & 4 \end{array}$$

$$G = \begin{array}{c|cc} & 1 & 2 \\ \hline 1 & 2 & 6 \\ 2 & 5 & 1 \end{array}$$

4. AC
5. $C_t A$
6. $C_t(AC)$ and $(C_t A)C$

where

$$A = \begin{array}{c|ccc} & 1 & 2 & 3 \\ \hline 1 & 2 & 0 & 4 \\ 2 & 0 & 5 & 1 \\ 3 & 7 & 3 & 6 \end{array}$$

$$C = \begin{array}{c|cc} & 1 & 2 \\ \hline 1 & 1 & 0 \\ 2 & 0 & 1 \\ 3 & 4 & 3 \end{array}$$

8. Let $B = \begin{array}{c|cc} & 1 & 2 \\ \hline 1 & 4 & 1 \\ 2 & 2 & 3 \end{array}$

Find D = symmetric part of B.

9. Find H = skew-symmetric part of B.
10. Calculate PBP
    PDP
    PHP

where

$$P = \begin{array}{c|c} & 1 \\ \hline 1 & 1 \\ 2 & 5 \end{array}$$

Problem 3.3

The generator and load currents and bus voltages for a given operating condition of the system of Figure 3.10 are

$$E_{G1} = 1.04 + j0.15 \qquad i_{G1} = 1.183 + j0.070$$
$$E_{G2} = 1.00 + j0.049 \qquad i_{G2} = 0.922 - j0.070$$
$$E_{G3} = 0.977 - j0.005 \qquad i_{G3} = 0 + j0$$
$$E_{L1} = 1.03 + j0.114 \qquad i_{L1} = -0.732 - j0.098$$
$$E_{L2} = 0.927 - j0.085 \qquad i_{L2} = -1.373 + j0.098$$

The impedance matrix with generator $G3$ as reference is given in equation 3–13.

From the above data determine the arithmetic value of the frame 2 impedances and voltages as denoted by equation 3–42. It is suggested that these quantities be calculated by using the transformation matrices indicated in equations 3–22, 3–23, and 3–24.

Problem 3.4

From the results of problem 3.3 determine the arithmetic value of reference frame 3 impedances and voltages as denoted by equation 3–60. It is suggested that the frame 3 quantities be calculated by using the transformation matrices indicated in equations 3–22, 3–23, and 3–24.

Problem 3.5

From problem 3.4 we have obtained

$$Z_{m-n} = \begin{array}{|c|c|c|} \hline 0.0671 + j0.1711 & 0.0041 - j0.0396 & 0.0041 - j0.0396 \\ \hline -0.0211 - j0.0304 & 0.0479 + j0.1189 & 0.0259 + j0.0589 \\ \hline -0.0211 - j0.0304 & 0.0259 + j0.0589 & 0.0259 + j0.0589 \\ \hline \end{array}$$

Calculate the symmetrical resistance matrix

$$R_{m-n} = \begin{array}{|c|c|c|} \hline & & \\ \hline & & \\ \hline & & \\ \hline \end{array}$$

Also calculate the skew-symmetric reactance matrix

$$\frac{(X_{m-n} - X_{n-m})}{2} = \begin{array}{|c|c|c|} \hline & & \\ \hline & & \\ \hline & & \\ \hline \end{array}$$

DEVELOPMENT OF TRANSMISSION LOSS FORMULA

Tabulate the losses obtained from

$$i_{dm}(R_{m-n})i_{dn} =$$
$$i_{qm}(R_{m-n})i_{qn} =$$
$$-2i_{dm}\frac{(+X_{m-n} - X_{n-m})}{2}i_{qn} =$$
$$\overline{}$$
$$P_L =$$

The above losses should check those obtained by $\sum i_k^2 R_k = 0.115$. The values of generator currents are obtained from problem 3.3 as

$$i_{d1} = 1.183 \qquad i_{q1} = +0.07$$
$$i_{d2} = 0.922 \qquad i_{q2} = -0.07$$
$$i_{d3} = 0 \qquad i_{q3} = 0$$

Problem 3.6

Consider the transmission systems given in Figure 3.30.

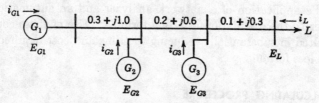

Figure 3.30. Transmission system for problem 3.6.

Recall that the reference frame 3 equations are given by

				G1	G2	G3		
G1	$E_{G1} - E_L$		G1	Z_{1-1}	Z_{1-2}	Z_{1-3}	G1	i_{G1}
G2	$E_{G2} - E_L$	=	G2	Z_{2-1}	Z_{2-2}	Z_{2-3}	G2	i_{G2}
G3	$E_{G3} - E_L$		G3	Z_{3-1}	Z_{3-2}	Z_{3-3}	G3	i_{G3}

Find the numerical value of the reference frame 3 impedances.

4 THE PRACTICAL CALCULATION OF LOSS FORMULA COEFFICIENTS

4.1 INTRODUCTION

This chapter is intended to provide a simple exposition [1] of the calculating procedure involved in determining a loss formula. It also discusses the application of a network analyzer and an automatic digital computer to the calculation of a loss formula. Use of these computers has resulted in substantially reducing the time and cost required to calculate a loss formula.

4.2 CALCULATING PROCEDURE

The simple three-generator system of Figure 4.1 is treated in detail.

A. Open-Circuit Impedance Measurements (Reference Frame 1)

The open-circuit impedance measurements with all loads, generators, synchronous condensers, and line-charging capacitors removed from ground are usually made first so that the matrices R_{Gm-Gn} and R_{Lj-Gn} may be calculated while the loads are being calibrated and the base case balanced. For the simplified system under consideration the matrices R_{Gm-Gn} and R_{Lj-Gn}, with Fort Wayne as reference, can be determined from the impedance diagram Figure 4.2 as

		G1	G2	G3	
Twin Branch	G1	0.0600	0	0	
R_{Gm-Gn} = Fort Wayne	G2	0	0	0	(4-1)
Portsmouth	G3	0	0	0.3209	

CALCULATION OF LOSS FORMULA COEFFICIENTS

$$R_{Lj-Gn} = \begin{array}{c} \\ \text{Twin Branch} \\ \text{Fort Wayne} \\ \text{Portsmouth} \\ \text{Muncie} \\ \text{Marion} \end{array} \begin{array}{c} \\ L1 \\ L2 \\ L3 \\ L4 \\ L5 \end{array}$$

		G1	G2	G3
Twin Branch	L1	0.0600	0	0
Fort Wayne	L2	0	0	0
Portsmouth	L3	0	0	0.3209
Muncie	L4	0	0	0.1089
Marion	L5	0	0	0.0771

(4-2)

The above resistances are on a 200-mva base.

In general, the reference bus, in this case Fort Wayne, is a major source chosen so that as many zero elements as possible will appear in the measured matrices. Each column of the matrices is determined in the following manner:

With the reference bus grounded, impress at Twin Branch a known current and determine the voltages appearing at all other load and generator buses. The real part of the ratio (bus voltage)/(known impressed current) is the desired resistance. Thus, if $1 + j0$ current is impressed at Twin Branch, $Z_{G1-G1} = (0.060 + j0.206)/(1 + j0) = 0.060 + j0.206$ and $R_{G1-G1} = 0.060$. Similarly, the voltages appearing at Fort Wayne, Portsmouth, Muncie, and Marion are zero; hence the remaining elements in the Twin Branch column are zero. When current is impressed at Fort Wayne all voltages are zero; therefore, all elements in column two are zero. The last column is determined when current is impressed from the last generator, Portsmouth. Complete results with current impressed at this generator are given in Figure 4.3.

The procedure described may be used directly on the electrical model representation provided by a network analyzer.

Methods have recently been developed of obtaining the self and mutual resistances of the network through means of an automatic digital computer.[2] These digital methods offer several distinct advantages over analogue methods:

1. Greater accuracy.
2. Lower cost.
3. Set-up time is small, as general programming decks are available.
4. Elimination of necessity of transcribing network-analyzer results to punched cards and associated time and possibility of error.

These methods are discussed in Section 4.10.

118 ECONOMIC OPERATION OF POWER SYSTEMS

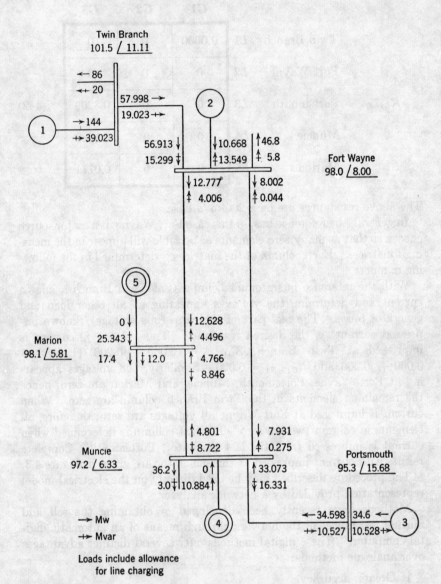

Figure 4.1. Base case load flow.

CALCULATION OF LOSS FORMULA COEFFICIENTS

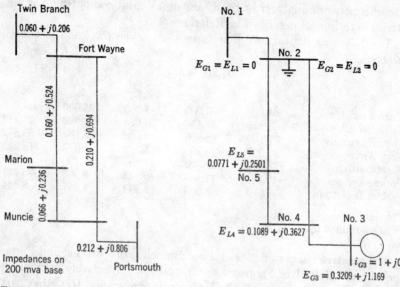

Figure 4.2. Impedance diagram.

Figure 4.3. Measurement of self and mutual resistances.

B. Normal Load Data

A typical operating condition in the range of load that is of interest is chosen as the base case to be set up. For this base case the following data are taken:

1. Record generator voltage magnitudes, megawatts, megavars, and angles.
2. Record scalar value of all line currents.
3. With the reference angle chosen so that the summation of the vector equivalent load equals a real number, record the real and imaginary parts of all equivalent load currents. The equivalent load current at a bus is defined as the sum of the synchronous condenser, line-charging, and load current at that bus. The load current at a source bus is to include that portion of the plant reactive which is not a function of generator output as discussed in Section 3.11.
4. For a check on the system operation it is desirable to record load voltages, load and line megawatts and megavars, and synchronous-condenser megavars.

It is suggested that a number of other load flows be taken to provide a check on the accuracy of the loss formula and to determine the manner in which the reactive power output of a given plant is related to the real power output of that plant. These load flows may be obtained by the use of a network analyzer or by means of a digital computer using the iterative technique described in Reference 3.

The following digital load-flow cases were taken for this illustrative problem:

TABLE 4.1

	1	2	3 *	4	5
Twin Branch mw	43.800	93.000	144.000	195.000	244.000
Twin Branch mvar	68.425	52.456	39.023	28.629	21.412
Fort Wayne mw	77.100	45.596	10.668	−22.162	−53.872
Fort Wayne mvar	−32.396	−25.290	−13.549	1.077	18.249
Portsmouth mw	73.500	51.500	34.600	18.200	4.900
Portsmouth mvar	−10.768	−11.468	−10.528	−8.369	−5.691
System Loss † mw	7.994	3.693	2.868	4.637	8.624

* Base case.
† Determined by $\sum I_k^2 R_k$

The reactive characteristics of Twin Branch, Fort Wayne, and Portsmouth, are plotted in Figures 4.4, 4.5, and 4.6, respectively. From an examination of these plots the following values of s_m and Q_{Lm}, that part of the plant reactive which is included with the load, are chosen:

$$\begin{array}{lll} \text{Twin Branch} & s_1 = -0.25 & Q_{L1} = +78 \text{ mvar} \\ \text{Fort Wayne} & s_2 = -0.40 & Q_{L2} = -6 \text{ mvar} \\ \text{Portsmouth} & s_3 = 0 & Q_{L3} = -9.5 \text{ mvar} \end{array}$$

C. Calculation of d_n

It has been shown in equation 3–90 that each

$$d_n = l_j' R_{Lj-Gn} = \frac{i_{Lj}'}{i_L} R_{Lj-Gn} = \frac{1}{i_L}(i_{Lj}' R_{Lj-Gn}) \qquad (4\text{--}3)$$

Thus

$$d_1 = l_j' R_{Lj-G1} = l_1' R_{L1-G1} + l_2' R_{L2-G1} + l_3' R_{L3-G1} + \cdots$$

where

$$l_j' = \mathcal{R}\frac{i_{Lj}}{i_L} \qquad (4\text{--}4)$$

$\mathcal{R}$ = real part of

i_{Lj} = load current at bus Lj from base case = $i_{Lj}' + ji_{Lj}''$

and $\quad i_L = \sum_j i_{Lj}$

CALCULATION OF LOSS FORMULA COEFFICIENTS

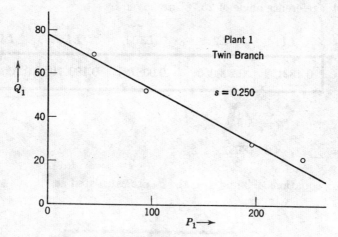

Figure 4.4. Twin Branch reactive characteristic.

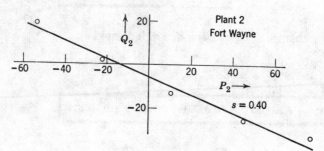

Figure 4.5. Fort Wayne reactive characteristic.

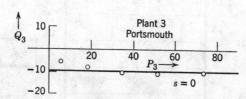

Figure 4.6. Portsmouth reactive characteristic.

ECONOMIC OPERATION OF POWER SYSTEMS

From the base case, the real equivalent load currents in p.u. projected against a reference angle of 25.78° are given by

$$i_{Lj}' = \begin{array}{|c|c|c|c|c|} \hline L1 & L2 & L3 & L4 & L5 \\ \hline 0.482158 & 0.209055 & -0.008721 & 0.189079 & 0.106511 \\ \hline \end{array}$$

(4-5)

Also

$$i_L = \sum_j i_{Lj} = 0.978082$$

Using equations 4-3 and 4-4, the d_n are calculated as

$$d_n = \begin{array}{|c|c|c|} \hline G1 & G2 & G3 \\ \hline 0.029578 & 0 & 0.026587 \\ \hline \end{array}$$

(4-6)

The mechanics of calculating d_n are indicated as

L1	L2	L3	L4	L5
0.482158	0.209055	-0.008721	0.189079	0.106511

	G1	G2	G3
↓ L1	0.0600	0	0
L2	0	0	0
L3	0	0	0.3209
L4	0	0	0.1089
L5	0	0	0.0771

$$= \begin{array}{|c|c|c|} \hline G1 & G2 & G3 \\ \hline i_L d_1 & i_L d_2 & i_L d_3 \\ \hline \end{array}$$

Thus

$$i_L d_3 = 0 + 0 + (-0.008721)(0.3209) + (0.189079)(0.1089)$$
$$+ (0.106511)(0.0771)$$

$$d_3 = 0.026587$$

CALCULATION OF LOSS FORMULA COEFFICIENTS

D. Calculation of f_n

It has been shown by equation 3-93 that

$$f_n = l_j'' R_{Lj-Gn} \tag{4-7}$$

where $\quad l_j'' = g \dfrac{i_{Lj}}{i_L} \tag{4-8}$

Thus $\quad f_1 = l_1'' R_{L1-G1} + l_2'' R_{L2-G1} + l_3'' R_{L3-G1} + \cdots \tag{4-9}$

From the base case, the imaginary equivalent load currents in p.u. projected against a reference angle of 25.78° are given by

	L1	L2	L3	L4	L5
$i_{Lj}'' =$	0.169336	−0.130198	−0.049074	−0.023723	0.033660

(4-10)

Note that $\quad \sum i_{Lj}'' = 0$

Using equations 4-7 and 4-8, the f_n are calculated as

	G1	G2	G3
$f_n =$	0.010388	0	−0.016089

(4-11)

The mechanics of calculating f_n are indicated as

→

L1	L2	L3	L4	L5
0.169336	−0.130198	−0.049074	−0.023723	0.033660

↓

	G1	G2	G3
L1	0.0600	0	0
L2	0	0	0
L3	0	0	0.3209
L4	0	0	0.1089
L5	0	0	0.0771

$=$

G1	G2	G3
$i_L f_1$	$i_L f_2$	$i_L f_3$

124 ECONOMIC OPERATION OF POWER SYSTEMS

Thus,

$$i_L f_3 = (0.169336)(0) - (0.130198)(0) - (0.049074)(0.3209)$$
$$- (0.023723)(0.1089) + (0.033660)(0.0771)$$
$$f_3 = -0.016089$$

E. Calculation of K_{mn}

From equation 3–104

$$K_{mn} = \frac{1}{V_m V_n}[(1 + s_m s_n)\cos\theta_{mn} + (s_m - s_n)\sin\theta_{mn}] \quad (4\text{-}12)$$

where $s_m = \dfrac{Q_m}{P_m}$ ratio for generator m as determined from curves or base case

θ_m = angle of generator m

$\theta_{mn} = \theta_m - \theta_n$

V_m = absolute value of voltage at generator m

Experience has shown that the loss formula is somewhat sensitive to the quantities s_m. Section 3.11 discusses in detail the considerations involved in the proper representation of source-reactive characteristics. It is usually worthwhile to obtain load-flow cases other than the base case to check the loss formula for different plant loadings. This additional data is also helpful in selecting the values of s_m.

The following data are tabulated from the base case and Figures 4.4, 4.5, and 4.6:

m	V_m	s_m	θ_m	Q_{Lm}
G1	1.015	−0.25	11.11	+78
G2	0.980	−0.40	8.00	−6
G3	0.953	0	15.68	−9.5

Q_{Lm} denotes that part of the plant reactive which is included with the load. It is convenient to calculate the K_{mn} in the following tabular form:

$m-n$	$V_m V_n$	θ_{mn}	$(1 + s_m s_n)\cos\theta_{mn}$	$(s_m - s_n)\sin\theta_{mn}$	K_{mn}
1−1	1.030225	0	1.062500	0	1.031328
1−2	0.994700	3.11	1.098383	0.008138	1.112414
1−3	0.967295	−4.57	0.996820	0.019920	1.051116
2−2	0.960400	0	1.160000	0	1.207830
2−3	0.933940	−7.68	0.991030	0.053456	1.118365
3−3	0.908209	0	1.000000	0	1.101068

CALCULATION OF LOSS FORMULA COEFFICIENTS

$$K_{mn} = \begin{array}{c|ccc} & G1 & G2 & G3 \\ \hline G1 & 1.031328 & 1.112414 & 1.051116 \\ G2 & 1.112414 & 1.207830 & 1.118365 \\ G3 & 1.051116 & 1.118365 & 1.101068 \end{array} \quad (4\text{-}13)$$

F. Calculation of H_{mn}

From equation 3-108

$$H_{mn} = \frac{1}{V_m V_n}[(1 + s_m s_n)\sin\theta_{mn} + (s_n - s_m)\cos\theta_{mn}] \quad (4\text{-}14)$$

The value of H_{mn} may be conveniently calculated as shown in the following table:

$m-n$	$V_m V_n$	θ_{mn}	$(1 + s_m s_n)\sin\theta_{mn}$	$(s_n - s_m)\cos\theta_{mn}$	H_{mn}
1-1	1.030225	0	0	0	0
1-2	0.994700	3.11	0.059675	-0.149780	-0.090581
1-3	0.967295	-4.57	-0.079680	0.249205	0.175260
2-2	0.960400	0	0	0	0
2-3	0.933940	-7.68	-0.133640	0.396412	0.281359
3-3	0.908209	0	0	0	0

The matrix of H_{mn} is tabulated as

$$H_{mn} = \begin{array}{c|ccc} & G1 & G2 & G3 \\ \hline G1 & 0 & -0.090581 & 0.175260 \\ G2 & 0.090581 & 0 & 0.281359 \\ G3 & -0.175260 & -0.281359 & 0 \end{array} \quad (4\text{-}15)$$

G. Calculation of A_{mn}, $P_m A_{mn} P_n$, and $P_m K_{mn} P_n$

From equation 3-115

$$w' = \frac{\sum i_k^2 R_k - P_m A_{mn} P_n + P_m H_{mn}(fm - f_n)P_n}{P_m K_{mn} P_n} \quad (4\text{-}16)$$

where $A_{mn} = (R_{Gm-Gn} - d_n - d_m)K_{mn} \quad (4\text{-}17)$

Then, for example

$$A_{23} = (R_{G2-G3} - d_2 - d_3)K_{23}$$
$$= (0 - 0 - 0.026587)(1.118365) = -0.029734$$

Also

$$A_{33} = (R_{G3-G3} - d_3 - d_3)K_{33}$$
$$= (0.3209 - 0.026587 - 0.026587)(1.101068) = 0.294785$$

Thus the matrix

	G1	G2	G3
G1	+0.000870	−0.032903	−0.059036
G2	−0.032903	0	−0.029734
G3	−0.059036	−0.029734	+0.294785

$A_{mn} = $ (above) (4-18)

From the base case, per unit,

	G1	G2	G3
$P_m = $	0.72000	0.05334	0.17300

A double summation is usually performed in two operations:

	G1	G2	G3			G1	G2	G3
$K_{mn}P_n = $	0.983735	1.058841	1.006942	=		$K_{1n}P_n$	$K_{2n}P_n$	$K_{3n}P_n$

For example

$$K_{1n}P_n = K_{11}P_1 + K_{12}P_2 + K_{13}P_3$$
$$= 1.031328(0.720000) + 1.112414(0.053340)$$
$$+ 1.051116(0.173000) = 0.983735$$

$$K_{2n}P_n = 1.112414(0.720000) + 1.207830(0.053340)$$
$$+ 1.118365(0.173000) = 1.058841$$

$$K_{3n}P_n = 1.051116(0.720000) + 1.118365(0.053340)$$
$$+ 1.101068(0.173000) = 1.006942$$

CALCULATION OF LOSS FORMULA COEFFICIENTS

so that

$$P_m K_{mn} P_n = 0.983735(0.720000) + 1.058841(0.053340) \\ + 1.006942(0.173000)$$

$$= 0.9389 \qquad (4\text{-}19)$$

Similarly,

$$A_{mn} P_n = \begin{array}{|c|c|c|} \hline G1 & G2 & G3 \\ \hline -0.011342 & -0.028834 & +0.006906 \\ \hline \end{array} \qquad (4\text{-}20)$$

and

$$P_m A_{mn} P_n = -0.008510$$

H. Calculation of $H_{mn}(f_m - f_n)$

The calculation of $H_{mn}(f_m - f_n)$ is illustrated by these sample calculations:

$$H_{12}(f_1 - f_2) = -0.090581(0.010388 - 0) = -0.000941$$

All the diagonal terms in which $m = n$ are, of course, identically zero.
Then

$$H_{mn}(f_m - f_n) = \begin{array}{c|c|c|c|} & G1 & G2 & G3 \\ \hline G1 & 0 & -0.000941 & 0.004640 \\ \hline G2 & -0.000941 & 0 & 0.004527 \\ \hline G3 & 0.004640 & 0.004527 & 0 \\ \hline \end{array} \qquad (4\text{-}21)$$

In a similar manner to that previously illustrated

$$P_m H_{mn}(f_m - f_n) P_n = 0.001167 \qquad (4\text{-}22)$$

I. Calculation of w'

Substituting in equation 4–16,

$$w' = \frac{0.014340 + 0.008510 + 0.001167}{0.9389} = 0.025579$$

where base case $I^2 R$ losses = 0.014340 p.u.

J. Determination of B_{mn}

From Section 3.10 and equation 3–113 we note that

$$B_{mn} = A_{mn} + w' K_{mn} - H_{mn}(f_m - f_n) \qquad (4\text{-}23)$$

For example

$$B_{11} = A_{11} + w'K_{11} - 0$$
$$= 0.000870 + (0.025579)(1.031328) = 0.027251$$
$$B_{23} = A_{23} + w'K_{23} - H_{23}(f_2 - f_3)$$
$$= -0.029734 + (0.025579)(1.118365)$$
$$- 0.004527 = -0.005653$$

and the matrix

	G1	G2	G3
G1	0.027251	−0.003506	−0.036788
$B_{mn} =$ G2	−0.003506	0.030896	−0.005653
G3	−0.036788	−0.005653	0.322950

(4–24)

An immediate check upon the solution for w' can be obtained by determining the losses with the B_{mn}. Using base case P_n,

	G1	G2	G3
$B_{mn}P_n =$	0.013069	−0.001854	0.029081

$P_L = P_m B_{mn} P_n = 0.014341$, which checks the base-case losses obtained by $\sum i_k^2 R_k$

K. Determination of Losses

If calculations of the form $P_m B_{mn} P_n$ are to be repeated with different P_m, labor can be saved by replacing B_{mn} with another matrix. It is shown in Section 3.8 that

$$P_m B_{mn} P_n = P_m D_{mn} P_n$$

where D_{mn} is a matrix such that

$$\frac{D_{mn} + D_{nm}}{2} = B_{mn}$$

CALCULATION OF LOSS FORMULA COEFFICIENTS

Thus, let

$$D_{mn} = \begin{array}{c|ccc} & G1 & G2 & G3 \\ \hline G1 & 0.027251 & -0.007012 & -0.073576 \\ G2 & 0 & 0.030896 & -0.011306 \\ G3 & 0 & 0 & 0.322950 \end{array} \quad (4\text{-}25)$$

It should be noted that the diagonal terms of D_{mn} are the same as B_{mn}, half of the off-diagonal terms is zero, and the other half is twice the corresponding B_{mn}.

For case 2,

$$P_n = \begin{array}{|c|c|c|} \hline G1 & G2 & G3 \\ \hline 0.720000 & 0.053340 & 0.173000 \\ \hline \end{array}$$

$$D_{mn}P_n = \begin{array}{|c|c|c|} \hline G1 & G2 & G3 \\ \hline 0.006518 & -0.000308 & 0.055870 \\ \hline \end{array}$$

$P_m D_{mn} P_n = 0.014341$

Similar calculations determine the losses for the remaining cases of Table 4.1. System losses, as determined from the loss formula and by summing the I^2R loss, line by line, are plotted against generation at Twin Branch in Figure 4.7.

4.3 EFFECT OF NEGLECTING PRODUCTS INVOLVING IMAGINARY LOAD CURRENTS

As indicated in Chapter 3, the calculation of a loss formula can be simplified if $P_m H_{mn}(f_m - f_n)P_n$ is negligibly small. Equation 4–16 then becomes

$$w' = \frac{\sum i_k^2 R_k - P_m A_{mn} P_n}{P_m K_{mn} P_n} \quad (4\text{-}26)$$

If $P_m H_{mn}(f_m - f_n)$ is neglected for the system studied in Section 4.2,

$$w' = \frac{0.014340 + 0.008510}{0.9389} = 0.024336$$

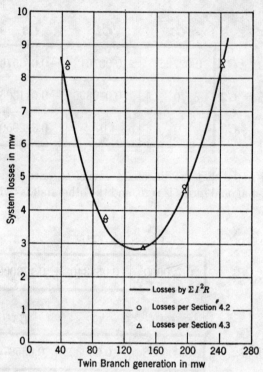

Figure 4.7. Comparison of losses by $\Sigma i_k^2 R_k$ and by loss formula.

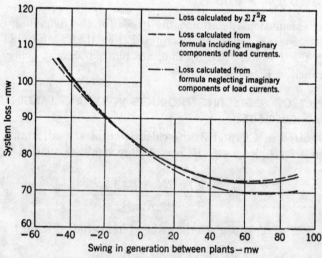

Figure 4.8. Improvement obtained by including effect of imaginary components of load currents.

CALCULATION OF LOSS FORMULA COEFFICIENTS

Also, equation 4-23 simplifies to

$$B_{mn} = A_{mn} + w'K_{mn} \qquad (4\text{-}27)$$

The new matrix of B_{mn} is given by

$$B_{mn} = \begin{array}{c|ccc} & G1 & G2 & G3 \\ \hline G1 & 0.025968 & -0.005831 & -0.033456 \\ G2 & -0.005831 & 0.029394 & -0.002517 \\ G3 & -0.033456 & -0.002517 & 0.321581 \end{array} \qquad (4\text{-}28)$$

Compare the B_{mn} matrix given by equation 4-28 with that previously given by equation 4-24. The results obtained with this loss formula are indicated by Δ in Figure 4.7.

It will be noted that for this particular problem excellent results are obtained with $P_m H_{mn}(f_m - f_n)P_n$ neglected.

Let us consider under what circumstances $P_m H_{mn}(f_m - f_n)P_n$ is small. From Chapter 3 we have

$$H_{mn} = \frac{1}{V_m V_n}[(1 + s_m s_n)\sin\theta_{mn} + (s_n - s_m)\cos\theta_{mn}] \qquad (4\text{-}29)$$

$$f_m - f_n = R_{Gm-Lk}l_k'' - R_{Gn-Lk}l_k'' \qquad (4\text{-}30)$$

The term H_{mn} is small if θ_{mn} is small and if $(s_n - s_m)$ is small. The term $f_m - f_n$ is usually small, since $\sum_k l_k'' = 0$. If all load voltages were at the same angle and all equivalent loads had the same power factor, f_m would be identically zero.

It has been found in many systems that the term $P_m H_{mn}(f_m - f_n)P_n$ may be assumed zero with negligible effect upon the accuracy of the loss formula. The term $P_m H_{mn}(f_m - f_n)P_n$ is usually found to be of value in systems in which the spread in generator angle across the system is large. Figure 4.8 is typical of the improvement obtained on a widespread system.

4.4 EFFECT OF APPROXIMATIONS IN REPRESENTATION OF PLANT REACTIVE CHARACTERISTICS

In this section we shall consider the effect of approximating the Twin Branch reactive characteristic given in Figure 4.4 by the relationship

$$Q_1 = s_1 P_1 \qquad (4\text{-}31)$$

If base-case data is used,

$$s_1 = \frac{Q_1}{P_1} = \frac{39.023}{144} = +0.271$$

The corresponding real load currents are then given by

	L1	L2	L3	L4	L5	
$i_{Lj} =$	0.433422	0.243238	0.011046	0.183434	0.085047	(4–32)

The values of d_n are given by

	G1	G2	G3	
$d_n =$	0.027197	0	0.031456	(4–33)

If the term $P_m H_{mn}(f_m - f_n)P_n$ is neglected, the resulting loss-formula coefficients are given by

		G1	G2	G3	
	G1	0.028652	−0.004941	−0.037061	
$B_{mn} =$	G2	−0.004941	0.026443	−0.010695	(4–34)
	G3	−0.037061	−0.010695	0.308168	

The corresponding losses are compared in Figure 4.9 with the losses obtained by $\sum i_k^2 R_k$. It will be noted that the shape of the curve has been somewhat distorted.

The results obtained with the approximation

$$Q_1 = 0.271 P_1$$

can be improved if s_1 is chosen to be a value more representative of higher plant loadings. For example, let us choose $s_1 = 0$. The corresponding loss-formula coefficients with $P_m H_{mn}(f_m - f_n)P_n$ neglected are

CALCULATION OF LOSS FORMULA COEFFICIENTS 133

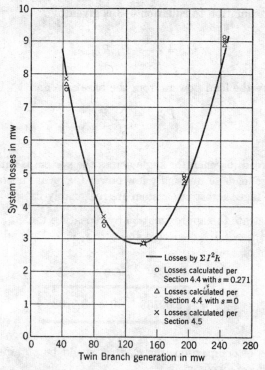

Figure 4.9. Comparison of losses by $\Sigma i_k{}^2 R_k$ and by loss formulas.

$$B_{mn} = \begin{array}{c|ccc} & G1 & G2 & G3 \\ \hline G1 & 0.028056 & -0.004000 & -0.036435 \\ G2 & -0.004000 & 0.028139 & -0.009124 \\ G3 & -0.036435 & -0.009124 & 0.309714 \end{array} \quad (4\text{-}35)$$

The losses obtained from the above coefficients are indicated by Δ in Figure 4.9.

4.5 APPROXIMATION OF d_n

From equations 4-3 and 4-4

$$d_n = \Re \frac{i_{Lj}}{i_L} R_{Lj-Gn} \qquad (4\text{-}36)$$

A close approximation to equation 4–3 is given by

$$d_n = \frac{1}{P_{\text{load}}} P_{Lj} R_{Lj-Gn} \qquad (4\text{--}37)$$

where P_{Lj} are the load powers from the base case and

$$P_{\text{load}} = \sum_j P_{Lj} \qquad (4\text{--}38)$$

provided that

1. The spread in generator angle across the system is small.
2. No large loads of unusually low power factor are uncorrected.
3. The voltages across the system are reasonably flat.

We shall illustrate the approximation by repeating the B_{mn} calculations with $P_m H_{mn}(f_m - f_n)P_n$ neglected and with $s_1 = -0.25$ and $Q_{L1} = +78$. The values of d_n are

	G1	G2	G3
$d_n =$	0.02768	0	0.02835

$(4\text{--}39)$

For purposes of comparison, the d_n that would be obtained by use of formula 4–36 are given as

	G1	G2	G3
$d_n =$	0.029578	0	0.026587

$(4\text{--}6)$

The resulting B_{mn} matrix is given by

		G1	G2	G3
	G1	0.027605	−0.006178	−0.035636
$B_{mn} =$	G2	−0.006178	0.026726	−0.006960
	G3	−0.035636	−0.006960	0.315265

$(4\text{--}40)$

The corresponding losses are indicated in Figure 4.9 by X.

4.6 DISCREPANCIES INVOLVED IN APPLICATION OF LOSS FORMULAS—STUDY OF SIMPLIFIED REPRESENTATION OF AGE SYSTEM

The discrepancies involved in the application of a given loss formula result chiefly from

1. Change in equivalent load-current pattern.
2. Change in assumed generator reactive characteristics.
3. Change in generator angular positions.
4. Change in generator-bus voltage magnitudes.

The effect of these factors has been investigated in a study of the American Gas and Electric System.[4] A simplified representation was used, maintaining, however, the wide variation of operating conditions existing in actual practice.

CHANGE IN LOAD PATTERN

In actual hourly operation the individual kilowatt loads of three operating divisions of the American Gas and Electric Company do not remain a fixed per cent of the total combined kilowatt load. The variation in the loads represented in this study are shown in this tabulation:

PER CENT VARIATION IN KILOWATT LOAD

	Loading Period				
	A	B	C	D	E
Indiana Division	25	22	22	19	19
Ohio Division	39	41	43	42	44
Appalachian Division	36	37	35	39	37

The various loading periods are illustrated in Figure 4.10. Results obtained in Reference 4 indicated that the effect of the variation in kilowatt load during the daily load cycle can be closely approximated by assuming that the equivalent load currents vary proportionally.

However, in the case of coal and steel strikes it was found that the load pattern may be changed considerably, and revised loss formulas were required.

CHANGE IN GENERATOR REACTIVE CHARACTERISTICS

In this particular study the generator reactive characteristics were represented by

$$Q_n = s_n P_n \qquad (4\text{--}41)$$

Improved accuracy was obtained by the relationship

$$Q_n = Q_{Ln} + s_n P_n \qquad (4\text{--}42)$$

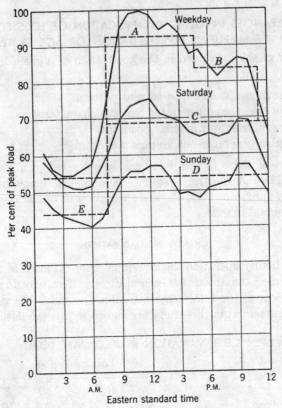

Figure 4.10. Daily load cycles.

VARIATION OF VOLTAGES

The results of this study indicated that the variations of generator voltage magnitudes and angles over the daily load cycle have a second-order effect on the accuracy of the loss formula.

USE OF AVERAGE LOSS FORMULA

For this particular study it was found that the loss formula obtained by averaging the B and D loading formulas was applicable for a normal variation in load pattern, as during the daily load cycle, and normal variations in generator-bus voltages and angles and moderate variations in generator Q/P ratios.

4.7 USE OF AN AUTOMATIC DIGITAL COMPUTING MACHINE

The methods described have been applied for example to the twenty-one variable system given in Figure 4.11. For formulas of this size it

CALCULATION OF LOSS FORMULA COEFFICIENTS 137

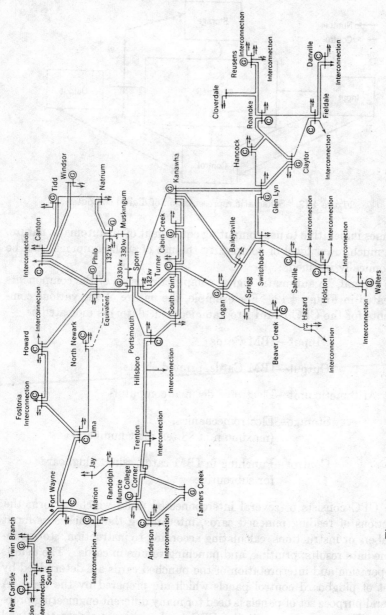

Figure 4.11. One-line diagram of American Gas and Electric System (1953).

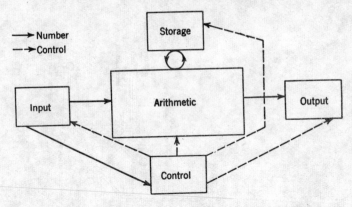

Figure 4.12. Schematic representation of digital computer.

becomes imperative to use computing equipment of an automatic nature. At a much lower level of complexity the use of such equipment can be economically justified.

In general, an automatic digital computer consists of the components indicated in Figure 4.12. For example, the nature of the various components for the CPC (Card Programmed Calculator) is indicated as,

 Input—IBM Cards

 Output—IBM Cards, printed sheets

 Arithmetic unit—Electronic decimal computer

  Storage—Electromechanical
   (maximum of 88 ten-digit numbers)

  Control—Punching in IBM cards, wired plugboards
   for subroutines

The CPC consists of several interconnected units which perform the functions of reading punched cards, interpreting the punches as input numbers or instructions, calculating according to instruction, storing intermediate results, printing, and punching results in cards. The details of operation and interpretation of the punched cards are determined by a set of plugboard control panels which are prepared by the user. A general purpose set of panels is used for many different engineering problems. These panels permit the CPC to do arithmetic operations and evaluate certain transcendental functions. Because the general purpose panels are used in the transmission-loss-formula calculation there is no machine set-up time to be considered.

As the machine accepts one or two instructions per card, programming the loss-formula calculation consists of breaking down the calculation into elemental arithmetic steps. They must then be described in codes for the machine. For example, consider the operation $A \times B = C$. The coding would be A code = 11, B code = 34, C code = 41, and operation code 3. The quantities A and B at this point appear in storage locations No. 11 and No. 34, respectively; the result C is to be stored in location No. 41; and the code for multiplication operation is 3. In this manner a deck of cards, describing the transmission-loss-formula calculation in its entirety, is assembled.

In a transmission-loss-formula study utilizing a network analyzer the first step after the analyzer plugging of the system is completed is to measure R_{Lj-Gn} and R_{Gm-Gn}. These measurements for a system such as Figure 4.11 would require approximately four hours. These impedances of the system may then be punched on special loading cards on which the data enters the calculator. The punching time involved for the impedance data of the system of Figure 4.11 would require approximately one and one half hours. While this is being done the base case is set up and recorded from the network analyzer; following this the base-case quantities i_{Lj}', i_{Lj}'', P_m, θ_m, V_m, s_m, and $\sum I^2 R$ are punched on loading cards. The punching of the base-case quantities for this system would require approximately one half hour.

The loading cards are collated with the deck of coded instruction cards, and the combined deck is placed in the machine. The calculation proceeds in a systematic manner as indicated by the computer flow diagram in Figure 4.13. The computer flow diagram, in general, indicates the calculating steps and logical steps which the computer follows. The calculations within a given step consisting of many numerical operations may be represented by a single box on the flow diagram. A logical step is one in which there is a choice involved. In a digital computer all choices are reduced to taking alternative instructions on negative values of certain numbers or to taking alternative instructions on zero values of certain numbers. A flow diagram is thus made up by connecting the arithmetic steps and logic steps with lines to indicate the sequence of operations. The flow diagram for the loss-formula calculation is very straightforward as the calculations proceed in a serial manner.

After the matrix is printed and punched in cards, the B_{mn} cards may be used in determination of system losses or penalty factors. The system of Figure 4.11, for which a twenty-four-generator programming deck was used, required less than eight hours running time on the CPC. For demonstration purposes the example of Figure 4.1 was run on the CPC with an instruction deck which caused intermediate as well as the final

ECONOMIC OPERATION OF POWER SYSTEMS

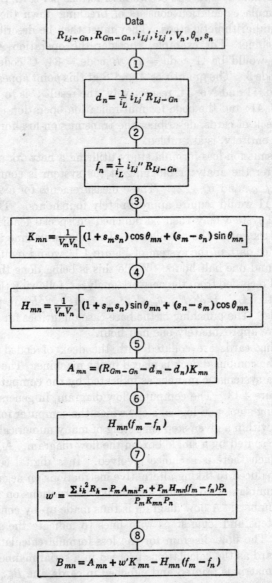

Figure 4.13. Computer flow diagram for loss formula calculation.

CALCULATION OF LOSS FORMULA COEFFICIENTS 141

results to be printed. The approximate time taken to process this example in the CPC using the method of Section 4.2 is two minutes. For fourteen generators the running time is less than four hours.

4.8 HIGH-SPEED INTERNALLY PROGRAMMED DIGITAL COMPUTER

The IBM Type 650 magnetic-drum data-processing machine is a high-speed, internally programmed, digital computer which has proved of great value in the calculation of transmission loss formulas. Compared to the CPC, the use of the 650 computer reduces the digital calculating time by approximately one fourth and the digital calculating cost by approximately one half. As an example of the form of the output, the results for the 650 computer calculation of the loss formula described in Section 4.2 are given in Figure 4.14. The 650 computer installation of the General Electric Analytical Engineering Section is shown in Figure 4.15.

The main data storage for the Type 650 computer is a magnetic drum which rotates at 12,000 rpm. Instructions and constants used in a particular problem are read into this drum initially and called upon when desired. On this drum are arranged storage positions for 2000 ten-digit numbers with their sign. The average time in which a number may be obtained from the drum is 2.4 milliseconds. The approximate operating times are

Multiplication	12.5 milliseconds
Division	16.5 milliseconds
Addition	3.5 milliseconds

The communication with the computer is in terms of standard punched cards with the input reading at a rate of 200 cards per minute and the output punching at the rate of 100 cards per minute. The instructions for this computer are of the single data address form. The ten-digit instruction word is composed of a two-digit operation instruction (add, subtract, divide, multiply, shift), a four-digit indication of the location in memory where the data is stored (data address), and a four-digit indication of where in memory the next instruction will be found.

For example, the computer instructions for the operation $(a)(b) = c$ would be written as

$a = 12345.67891$	at address 1001
$b = 023.4567891$	at address 1002
$c = 289539.9649$	to be sent to 1003

142 ECONOMIC OPERATION OF POWER SYSTEMS

238	2	.978082	.000001			$\sum I_{Lj}'$ $\sum I_{Lj}''$
567	5	.482158	.209055	.008721−	.189079	.106511	I_{Lj}'
687	5	.169336	.130198−	.049074−	.023723−	.033660	I_{Lj}''
493	3	.029578	.000000	.026587			d_n
530	3	.010388	.000000	.016089−			f_n
1234	6	1.031328	1.112414	1.051116	1.207830	1.118365	1.101068	K_{mn}
567	6	.000000	.090581−	.175260	.000000	.281359	.000000	H_{mn}
567	6	.000870	.031962−	.063676−	.000000	.034261−	.294785	A_{mn}
		.9389	.008510−	.014340			$P_m K_{mn} P_n$ $P_m A_{mn} P_n$ I_n^2
1234	6	.027251	.003506−	.036788−	.030896	.005653−	.322950	B_{mn}
1		.014342					$P_m B_{mn} P_n$ (base case)

Figure 4.14. Tabulation of results from automatic digital computer.

CALCULATION OF LOSS FORMULA COEFFICIENTS

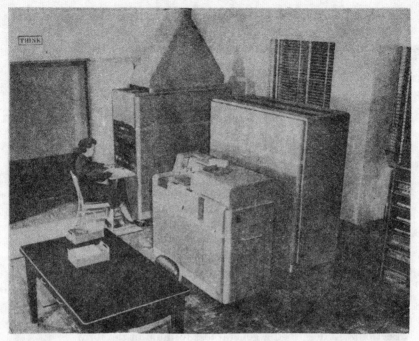

Figure 4.15. General Electric Analytical Engineering Section IBM 650 computer installation.

	Instruction			
Location of Instruction	Operation	Data Address	Next Instruction Address	Description
0500	60	1001	0501	Reset add a
0501	19	1002	0502	Multiply by b
0502	31	0008	0503	Shift right 8 and round
0503	20	1003	0504	Store result c

This type of computer is so balanced that it is of approximately equal value for financial computations involving large volumes of input-output data and for engineering calculations which usually involve a moderate amount of data and a fairly long calculating sequence.

Larger scale computers, such as the IBM 704 Electronic Data-Processing Machine, offer additional economies of calculation concerning the determination of digital load flows [5] and transmission loss formulas. A computer of this magnitude allows the complete determination of load flows and loss formulas without intermediate manual data handling.

The internal high-speed storage of the 704 consists of magnetic cores. Using this high-speed storage, which has an access time of 12 microseconds, the following operating times are obtained for floating point operations:

Multiplication	204 microseconds
Division	216 microseconds
Addition	84 microseconds

The IBM 704 computer is available with 4096, 8192, or 32,768 locations of high-speed storage. For greater internal working storage, as well as their input-output function, magnetic tape units capable of transcribing 2500 eight-digit decimal numbers per second are available. Other input-output components are the card reader and card punch and the direct connected printer. Reference 5 estimates that the 704 computer results in excess of a 3-to-1 cost advantage over the 650 for calculating load flows. Experience has indicated that the 704 computer can produce the load flows required for loss formulas at costs comparable to the network analyzer.

4.9 CALCULATION OF GENERAL LOSS FORMULA

Section 3.12 described a general loss formula of the form

$$P_L = P_m B_{mn} P_n + B_{no} P_n + B_{oo} \qquad (3\text{-}120)$$

and indicated that the calculation of B_{no} and B_{oo} by equations 3-122 and 3-126 would be quite lengthy, since these equations require the determination of the self and mutual impedances between loads. This section describes a method [6] of calculating the B_{no} and B_{oo} terms which does not require the determination of the self and mutual impedances between loads. The method discussed involves the use of circuit theory together with a least-squares solution for w', B_{no}, and B_{oo}. The philosophy behind this method involves the concept of using w', B_{no}, and B_{oo} to fit the formula to observed transmission-loss data.

To understand the method, we will first briefly review the idea of multiple correlation with a least-squares criterion.[7,8] Consider the following function for which y is defined as

$$y = a_1 x_1 + a_2 x_2 \qquad (4\text{-}43)$$

It is desired to determine a_1 and a_2 when given observed values of y for various values of x_1 and x_2 such that the summation of the squares of the residuals is a minimum. A residual is determined as the deviation

CALCULATION OF LOSS FORMULA COEFFICIENTS

of the computed value from the observed value. Thus, for case 1 the residual v is given by

$$v_1 = a_1 x_1^{(1)} + a_2 x_2^{(1)} - y_1 \qquad (4\text{-}44)$$

where $\quad y_1 =$ observed value for case 1

$x_1^{(1)} =$ value of x_1 for case 1

$x_2^{(1)} =$ value of x_2 for case 1

Assuming that we have four cases, we have

$$v_2 = a_1 x_1^{(2)} + a_2 x_2^{(2)} - y_2 \qquad (4\text{-}45)$$

$$v_3 = a_1 x_1^{(3)} + a_2 x_2^{(3)} - y_3 \qquad (4\text{-}46)$$

$$v_4 = a_1 x_1^{(4)} + a_2 x_2^{(4)} - y_4 \qquad (4\text{-}47)$$

The method of least squares results in a solution for the desired coefficients such that the sum of the squares of the residuals is a minimum. Thus, it is desired that

$$S = \text{sum of squares of residuals}$$
$$= v_1^2 + v_2^2 + v_3^2 + v_4^2 = \text{minimum} \qquad (4\text{-}48)$$

A necessary condition that $S =$ minimum is that

$$\frac{\partial S}{\partial a_1} = 0 \qquad (4\text{-}49)$$

$$\frac{\partial S}{\partial a_2} = 0 \qquad (4\text{-}50)$$

From equation 4–48 we have

$$\frac{\partial S}{\partial a_1} = 2 \frac{\partial v_1}{\partial a_1} v_1 + 2 \frac{\partial v_2}{\partial a_1} v_2$$
$$+ 2 \frac{\partial v_3}{\partial a_1} v_3 + 2 \frac{\partial v_4}{\partial a_1} v_4 \qquad (4\text{-}51)$$

$$\frac{\partial S}{\partial a_2} = 2 \frac{\partial v_1}{\partial a_2} v_1 + 2 \frac{\partial v_2}{\partial a_2} v_2$$
$$+ 2 \frac{\partial v_3}{\partial a_2} v_3 + 2 \frac{\partial v_4}{\partial a_2} v_4 \qquad (4\text{-}52)$$

From equations 4-44 to 4-47 we note

$$\frac{\partial v_1}{\partial a_1} = x_1^{(1)} \qquad (4\text{-}53)$$

$$\frac{\partial v_2}{\partial a_1} = x_1^{(2)} \qquad (4\text{-}54)$$

$$\frac{\partial v_3}{\partial a_1} = x_1^{(3)} \qquad (4\text{-}55)$$

$$\frac{\partial v_4}{\partial a_1} = x_1^{(4)} \qquad (4\text{-}56)$$

Substituting equations 4-53 to 4-56 into equation 4-51, we obtain

$$\frac{\partial S}{\partial a_1} = 2x_1^{(1)}v_1 + 2x_1^{(2)}v_2 + 2x_1^{(3)}v_3 + 2x_1^{(4)}v_4 = 0 \quad (4\text{-}57)$$

Similarly,

$$\frac{\partial S}{\partial a_2} = 2x_2^{(1)}v_1 + 2x_2^{(2)}v_2 + 2x_2^{(3)}v_3 + 2x_2^{(4)}v_4 = 0 \quad (4\text{-}58)$$

Substituting the definitions for v_1, v_2, v_3, and v_4 into equation 4-57, we obtain

$$\frac{\partial S}{\partial a_1} = 2x_1^{(1)}[a_1 x_1^{(1)} + a_2 x_2^{(1)} - y_1] + 2x_1^{(2)}[a_1 x_1^{(2)} + a_2 x_2^{(2)} - y_2]$$

$$+ 2x_1^{(3)}[a_1 x_1^{(3)} + a_2 x_2^{(3)} - y_3] + 2x_1^{(4)}[a_1 x_1^{(4)} + a_2 x_2^{(4)} - y_4]$$

$$(4\text{-}59)$$

Grouping the coefficients of a_1 and a_2 and dividing by (2), equation 4-59 becomes

$$a_1[x_1^{(1)}x_1^{(1)} + x_1^{(2)}x_1^{(2)} + x_1^{(3)}x_1^{(3)} + x_1^{(4)}x_1^{(4)}]$$

$$+ a_2[x_1^{(1)}x_2^{(1)} + x_1^{(2)}x_2^{(2)} + x_1^{(3)}x_2^{(3)} + x_1^{(4)}x_2^{(4)}]$$

$$= x_1^{(1)}y_1 + x_1^{(2)}y_2 + x_1^{(3)}y_3 + x_1^{(4)}y_4 \quad (4\text{-}60)$$

Similarly, equation 4-52 becomes

$$a_1[x_2^{(1)}x_1^{(1)} + x_2^{(2)}x_1^{(2)} + x_2^{(3)}x_1^{(3)} + x_2^{(4)}x_1^{(4)}]$$

$$+ a_2[x_2^{(1)}x_2^{(1)} + x_2^{(2)}x_2^{(2)} + x_2^{(3)}x_2^{(3)} + x_2^{(4)}x_2^{(4)}]$$

$$= x_2^{(1)}y_1 + x_2^{(2)}y_2 + x_2^{(3)}y_3 + x_2^{(4)}y_4 \quad (4\text{-}61)$$

CALCULATION OF LOSS FORMULA COEFFICIENTS

The values of a_1 and a_2 obtained from the solution of the normalized simultaneous equations 4–60 and 4–61 satisfy the necessary condition that the summation of squares of the residuals is a minimum.

For the general cases, define y as

$$y = \sum_{j=1}^{r} a_j x_j \tag{4-62}$$

for which there are r coefficients to be determined. Designate by n the number of observations. The summation of the squares of the residuals is given by

$$S = \sum_{i=1}^{n} v_i^2 \tag{4-63}$$

where

$$v_i = \sum_{j=1}^{r} a_j x_j^{(i)} - y_i \tag{4-64}$$

The necessary condition that S is a minimum is given by

$$\frac{\partial S}{\partial a_k} = 0 = 2 \sum_{i=1}^{n} \frac{\partial v_i}{\partial a_k} v_i \tag{4-65}$$

where $k = 1, 2, \cdots r$

From equation 4–64 we have

$$\frac{\partial v_i}{\partial a_k} = x_k^{(i)} \tag{4-66}$$

Substituting equations 4–66 and 4–64 into equation 4–65,

$$\frac{\partial S}{\partial a_k} = \sum_{i=1}^{n} \left[\sum_{j=1}^{r} a_j x_j^{(i)} - y_i \right] x_k^{(i)} = 0 \tag{4-67}$$

Collecting coefficients of a_j, we obtain the following normalized simultaneous equations to solve for the r unknown coefficients:

$$\sum_{j=1}^{r} \left[\sum_{i=1}^{n} x_j^{(i)} x_k^{(i)} \right] a_j = \sum_{i=1}^{n} y_i x_k^{(i)} \tag{4-68}$$

Having reviewed the method of least squares, let us consider the application of this method to the determination of the loss-formula coefficients.

From equations 3–120 and 3–113

$$P_L = P_m B_m P_n + B_{no} P_n + B_{oo} \tag{3-120}$$

$$= P_m A_{mn} P_n - P_m H_{mn}(f_m - f_n) P_n$$

$$+ P_m K_{mn} P_n w' + B_{no} P_n + B_{oo} \tag{4-69}$$

The quantities A_{mn} and $H_{mn}(f_m - f_n)$ may be easily calculated as indicated by equations 3–90, 3–93, 3–104, 3–108, and 3–114. Knowing P_L from load-flow data, we then write the driving function as

$$P_L - P_m A_{mn} P_n + P_n H_{mn}(f_m - f_n) P_n = w'(P_m K_{mn} P_n)$$
$$+ B_{no}(P_n) + B_{oo}(1) \quad (4\text{--}70)$$

From equation 4–70 it is desired to obtain w', B_{no}, and B_{oo} by the method of least squares. Since it is necessary to determine $(n + 2)$ unknowns, it is suggested that $2(n + 2)$ cases be considered in order to avoid distorting the loss formula to fit a small number of cases exactly. In comparing equation 4–70 to equation 4–62 let

$$y = P_L - P_m A_{mn} P_n + P_m H_{mn}(f_m - f_n) P_n \quad (4\text{--}71)$$

$$a_j = w', B_{no}, B_{oo} \quad (4\text{--}72)$$

$$x_j = P_m K_{mn} P_n, P_n, 1 \quad (4\text{--}73)$$

The solution of equation 4–70, according to the least-squares method, may be efficiently accomplished by the use of an automatic digital computer. In order to calculate a loss formula by these methods the computer flow design of Figure 4.13 is then modified to that shown in Figure 4.16. Steps 1, 2, and 7 have been changed. In steps 3 and 4 the values of s_m, θ_m, and V_m correspond to the average of the on-peak and off-peak data.

4.10 DIGITAL CALCULATION OF SELF AND MUTUAL IMPEDANCES

Reference 2 describes three digital methods of calculating self and mutual impedances on an automatic digital computer. Method 1 is an application of Gabriel Kron's circuit transformation for stationary networks [9] and involves straightforward matrix operations. Method 2 defines the network in terms of tracks and loops and determines the solution by superimposing a set of balancing currents upon an assumed set of currents. Both of these methods are direct in that the answers produced are accurate within the precision carried in the problem solution. Method 3 is an iterative nodal method in which an assumed set of voltages is successively improved until satisfaction of Kirchhoff's law has been achieved within the precision desired in the problem solution.

In Method 2, as previously stated, the network is defined in terms of loops and tracks. A procedural requirement is that loop currents threading a common branch do so in the same direction. Positive direction of flow in radial branches is assumed directed into the loop network. The current directions indicated in the illustrative system of Figure 4.17 have been chosen to satisfy these conventions.

CALCULATION OF LOSS FORMULA COEFFICIENTS

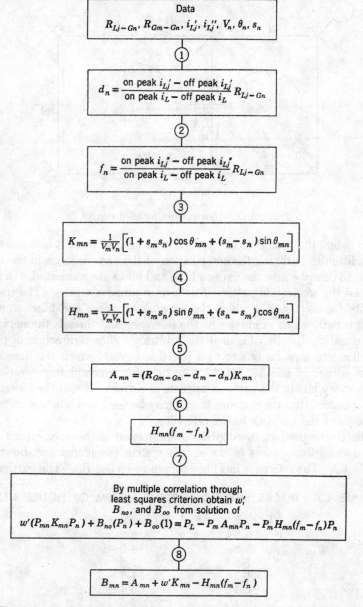

Figure 4.16. Computer flow diagram for general loss formula calculation.

150   ECONOMIC OPERATION OF POWER SYSTEMS

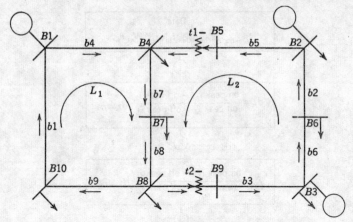

Figure 4.17. Illustrative simplified system.

To define the geometry or configuration of a network it is necessary, in addition to outlining the composition of the various loops in the network, to describe how the various lines and buses are connected. In this method the connection pattern is defined in terms of tracks. The tracks specify the sequence in which the network branches and buses appear along a path which originates at the reference and threads the network in a negative direction (against the arbitrarily chosen direction of positive flow) to a radial bus or to a junction beyond which the lines and buses appear in another track. The network is completely "tracked" when every bus in the network appears in a track. From the foregoing it is apparent that the network tracks may be used to relate bus voltages throughout the network to the reference.

The tracks used in the digital determination of the open-circuit self and mutual impedances of the simple system considered are shown in Table 4.2. The reference has been chosen to be bus $B3$. The particular

TABLE 4.2. TRACKS SELECTED FOR THE SYSTEM OF FIGURE 4.17

Track 1, $B - b$	Track 2, $B - b$
9–3	9–3
8–2–	8–2–
7–8	7–8
4–7	4–7
1–4	5–1–
10–1	2–5
8–9	6–2
	3–6

notational scheme used for this study is that an element in the track is specified by $B - b\pm$ where B is the bus at the input end of branch b,

CALCULATION OF LOSS FORMULA COEFFICIENTS

shown plus for lines and minus for transformers. For example, track 1 is indicated by a dashed-line path drawn in Figure 4.18.

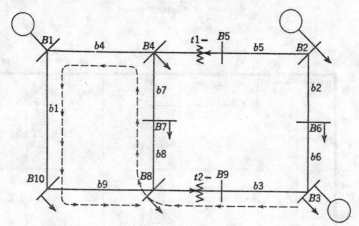

Figure 4.18. Simplified system with track 1 identified.

ASSUMED FLOWS AND RESULTING VOLTAGE DROPS DUE TO IMPRESSED CURRENTS

As pointed out in the statement of Method 2, it is necessary to assume a distribution of flow in the network due to impressing current at a given bus. The only requirement on the assumed flow distribution is that there be continuity of flow, i.e., that the current entering a junction equal that leaving the junction. From this and from the previous discussion of tracks it is seen that the assumed flow between the energized bus and the reference may be assumed through the line elements in the track between the energized bus and the reference. Of course, when off-nominal transformer ratios exist in the track their effect must be considered in determining the assumed flow in the elements between the transformer and the reference bus. The assumed branch flows resulting from impressing a current of $1 + j0$ at bus 1 are shown in Table 4.3 as I_{branch}.

TABLE 4.3. BRANCH FLOWS AND VOLTAGES DUE TO IMPRESSING $1 + j0$ AT BUS 1

Branch	I_{branch}	E_{branch}	I_{bal}	I_{total}	$Z_{\text{branch}} I_{\text{total}}$
1	0	0	$-0.30468 - j0.00419$	$-0.30468 - j0.00419$	$-0.01227 - j0.03857$
2	0	0	$-0.54547 + j0.01457$	$-0.54547 + j0.01457$	$-0.00271 - j0.01986$
3	$1.09 + j0$	$0.00654 + j0.15260$	$-0.55566 + j0.01484$	$0.53434 + j0.01484$	$+0.00113 + j0.07490$
4	$1 + j0$	$0.02800 + j0.08400$	$-0.30468 - j0.00419$	$0.69532 - j0.00419$	$+0.01982 + j0.05829$
5	0	0	$-0.54547 + j0.01457$	$-0.54547 + j0.01457$	$-0.00395 - j0.06621$
6	0	0	$-0.54547 + j0.01457$	$-0.54547 + j0.01457$	$-0.00271 - j0.01986$
7	$1 + j0$	$0.02800 + j0.08400$	$-0.81447 + j0.00942$	$0.18553 + j0.00942$	$+0.00440 + j0.01585$
8	$1 + j0$	$0.02800 + j0.08400$	$-0.81447 + j0.00942$	$0.18553 + j0.00942$	$+0.00440 + j0.01585$
9	0	0	$-0.30468 - j0.00419$	$-0.30468 - j0.00419$	$-0.01636 - j0.05142$

The branch voltages due to these current flows are shown as E_{branch} and are given by

$$E_{\text{branch}} = Z_{\text{branch}} I_{\text{branch}} \qquad (4\text{--}74)$$

where $Z_{\text{branch}} =$

branch	branch b1	b2	b3	b4	b5	b6	b7	b8	b9
b1	0.042+ j0.126								
b2		0.004+ j0.0365							
b3			0.006+ j0.140						
b4				0.028+ j0.084					
b5					0.004+ j0.1215				
b6						0.004+ j0.0365			
b7							0.028+ j0.084		
b8								0.028+ j0.084	
b9									0.056+ j0.168

$$(4\text{--}75)$$

The branch impedances are the line and transformer impedances of the network. The axis numbers are the numbers assigned to the elements in Figure 4.17.

LOOP-VOLTAGE DROPS AND BALANCING FLOWS

In this method of analysis the branch-current distribution resulting from current impressed at a given bus may be considered to have two components: (1) arbitrarily assumed distribution of current and (2) a set of balancing currents which circulate in the network loops. The balancing currents produce voltages in the network loops which nullify those produced by the assumed currents. Thus the superposition of the balancing flows on the assumed flows results in the correct distribution of flow in which each of the fundamental electric-circuit relationships are satisfied, i.e., (1) the summation of currents into a junction is zero, and (2) the summation of branch voltages around each loop is zero.

CALCULATION OF LOSS FORMULA COEFFICIENTS

The relationship between the set of loop-balancing currents and the set of loop voltages resulting from an assumed distribution of flow is

$$I_{\text{loop}} = -Y_{\text{loop}} E_{\text{loop}} \qquad (4\text{-}76)$$

Here Y_{loop} is the loop-admittance matrix and is equal to the inverse of the loop-impedance matrix. The loop-impedance matrix Z_{loop} may be obtained directly by the operation $C_t' Z_{\text{branch}} C'$ where C_t' is defined by

$$E_{\text{loop}} = C_t' E_{\text{branch}} \qquad (4\text{-}77)$$

In a network in which there are no off-nominal transformer ratios the net voltage acting around each loop is obtained merely by summing the branch voltages around the loop. However, when off-nominal ratios are present the net loop voltage is computed in terms of the branch voltages as viewed from an arbitrary reference point in the loop. Similarly, in a loop with no transformers the circulating current has the same value in each branch in the loop. When transformers are encountered the circulating current is expressed in terms of that value existing at the reference point in the loop.

For the system of Figure 4.17 the loop voltages E_{L1} and E_{L2} are related to the branch voltages $E_{b1}, E_{b2}, \cdots E_{b9}$ by the transformation matrix C_t' given below:

$$C_t' = \begin{array}{c|c|c|c|c|c|c|c|c|c} & b1 & b2 & b3 & b4 & b5 & b6 & b7 & b8 & b9 \\ \hline L1 & 1 & & & 1 & & & 1 & 1 & 1 \\ \hline L2 & & 1 & 1.01869 & & 1 & 1 & 0.93458 & 0.93458 & \end{array}$$

$$(4\text{-}78)$$

This matrix indicates that loop 1 is composed of branches 1, 4, 7, 8, and 9, and loop 2 is made up of branches 2, 3, 5, 6, 7, and 8. The effects of off-nominal turns ratios are also represented. For example, the voltage acting around loop 2, as viewed looking upward from bus 3, is equal to the voltages in branches 6, 2, and 5, plus the voltages in 7 and 8 multiplied by the turns ratio of transformer 1, plus the voltage in branch 3 multiplied by the product of the turns ratios of transformers 1 and 2. The loop-impedance matrix Z_{loop} is then given by $C_t' Z_{\text{branch}} C$ as

$$Z_{\text{loop}} = \begin{array}{c|c|c} & L1 & L2 \\ \hline L1 & 0.18200 + j0.54600 & 0.05234 + j0.15701 \\ \hline L2 & 0.05234 + j0.15701 & 0.06714 + j0.48652 \end{array} \qquad (4\text{-}79)$$

The network loop-admittance matrix may be obtained [9] by inverting Z_{loop}. Thus

$$Z_{\text{loop}}^{-1} = Y_{\text{loop}} = \begin{array}{c|c|c} & L_1 & L2 \\ \hline L1 & 0.57125 - j1.63313 & -0.07582 + j0.64257 \\ \hline L2 & -0.07582 + j0.64257 & 0.26372 - j2.23457 \end{array} \quad (4\text{-}80)$$

The procedure for calculating an inverse is given in Section 6.5.

The voltages in loops 1 and 2 resulting from the assumed flow distribution for current impressed at bus 1 are calculated from equation 4–77. Thus

$$E_{\text{loop}} = \begin{array}{c|c} L1 & 0.08400 + j0.25200 \\ \hline L2 & 0.05900 + j0.31246 \end{array} \quad (4\text{-}81)$$

For example,

$$E_{L1} = E_{b1} + E_{b4} + E_{b7} + E_{b8} + E_{b9}$$

$$= 0 + (0.02800 + j0.08400) + (0.02800 + j0.08400)$$
$$\quad + (0.02800 + j0.08400) + 0$$

$$= 0.08400 + j0.25200$$

Matrices 4–80 and 4–81 can now be used in equation 4–76 to determine the loop-balancing currents

$$I_{\text{loop}} = \begin{array}{c|c} L1 & -0.30468 - j0.00419 \\ \hline L2 & -0.54547 + j0.01457 \end{array} \quad (4\text{-}82)$$

BRANCH FLOWS FROM LOOP FLOWS

The balancing currents flowing in the individual branches of the network are related to the loop-circulating currents by the matrix C'. This

CALCULATION OF LOSS FORMULA COEFFICIENTS

is the transpose of the transformation matrix C_t' which in turn relates loop voltages to branch voltages.

Branch-balancing currents obtained by

$$I_{bal} = C'I_{loop} \tag{4-83}$$

for impressed current of $1 + j0$ at bus 1 are given in Table 4.3 as I_{bal}. The exact branch currents (the balancing currents superimposed on the assumed flows) are also shown in Table 4.3 as

$$I_{total} = I_{bal} + I_{branch} \tag{4-84}$$

BRANCH-VOLTAGE DROPS AND BUS VOLTAGES

The branch voltages due to the exact current distribution must next be computed. These voltages, of course, are related to the current flows by the branch impedances. Thus

$$E_{branch} = Z_{branch}I_{total} \tag{4-85}$$

The branch voltages resulting from unit current impressed st bus 1 are shown in Table 4.3.

The voltages at each network bus are then determined with respect to the reference bus by summing the branch voltages in each track away from the reference. Voltages determined in this fashion are the open-circuit impedances between the energized bus and the other network buses. The form of this calculation is indicated by Table 4.4 for the case in which bus 1 is energized with $1 + j0$ current.

TABLE 4.4. COMPUTER RESULTS OF METHOD 2, BUS 1 OF FIGURE 4.17 ENERGIZED

Bb	R	X
0903	0.001133	0.074898
0802−	0.001235	0.081639
0708	0.005641	0.097489
0407	0.010047	0.113339
0104	0.029866	0.171630
1001	0.017595	0.133067
0809	0.001234	0.081648
0903	0.001133	0.074898
0802−	0.001235	0.081639
0708	0.005641	0.097489
0407	0.010047	0.113339
0501−	0.009390	0.105924
0205	0.005441	0.039710
0602	0.002728	0.019859
0306	0.000015	0.000008

FLOW DIAGRAMS

Figure 4.19 shows a flow diagram for Method 2. This flow diagram indicates only the major steps, and what appears to be a simple arith-

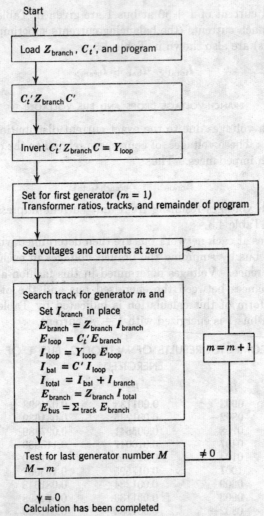

Figure 4.19. Computer flow diagram for Method 2.

metic box may, in turn, itself be represented by a detailed flow diagram. The first major step consists of loading into machine memory the branch impedances, the connection matrix C_t', and the program for determining

CALCULATION OF LOSS FORMULA COEFFICIENTS

Y_{loop}. Then the calculation of the matrices Z_{loop} and its inverse Y_{loop} is performed. Next, the generator index m is set equal to 1 and the remainder of the program and remaining data required are loaded. The foregoing has consisted of preparation for the main calculation which is to be repeated for each generator. This main calculation consists of initializing various voltages and currents by setting them equal to zero and then performing the arithmetic necessary to develop the set of bus voltages throughout the system. With the arithmetic performed and the answers punched out, a test is made to see if this calculation has been performed for the last generator or if more similar calculations remain to be done. If more calculations remain, the index m is increased by 1 and another pass through the main calculating loop is made. Eventually m becomes equal to M, the total number of generators, and the calculation is complete.

COMPUTER TIME

To assist in judging the relative merits of obtaining self and mutual impedances digitally by Method 2 or by network-analyzer tests, comparative calculations were undertaken for a system consisting of 30 buses, 38 lines, 14 loops, and 5 auto-transformers representing off-nominal turn ratios. Method 2 required approximately one hour of 650 computer time. The time to obtain and record these same results from the network analyzer is approximately four hours. Usually, in obtaining network-analyzer results, it is necessary to average the mutual impedances to obtain a perfectly symmetrical matrix. A digital approach is particularly advantageous when (1) high accuracy is required and (2) the self and mutual impedances are required in the form of punched cards.

4.11 SUMMARY

This chapter provides a simple exposition of the calculating procedure involved in determining transmission-loss formulas. The application of analogue and digital computers to these calculations is also described. The use of these computers has resulted in a substantial reduction in time and cost involved in obtaining a loss formula. With the increasing availability of high-speed, automatic digital computers, an increasingly greater portion of the required calculations will be accomplished by digital computers.

References

1. Loss Formulas Made Easy, A. F. Glimn, R. Habermann, Jr., L. K. Kirchmayer, G. W. Stagg. *AIEE Trans.*, Vol. 72, Part III, 1953, pp. 730–737.

2. Digital Calculation of Network Impedances, A. F. Glimn, R. Habermann, Jr., J. M. Henderson, L. K. Kirchmayer. *AIEE Trans.*, Vol. 74, Part III, 1955, pp. 1285–1296.
3. Automatic Calculation of Load Flows, A. F. Glimn, G. W. Stagg. *AIEE Trans. Paper 57-681* presented at the Summer Meeting, Montreal, Quebec, June 24–28, 1957.
4. Analysis of Total and Incremental Losses in Transmission Systems, L. K. Kirchmayer, G. W. Stagg. *AIEE Trans.*, Vol. 70, Part I, 1951, pp. 1197–1205.
5. Extensions in Digital Load-Flow Techniques Using High Speed Computer, M. S. Dyrkacz, A. F. Glimn, D. G. Lewis. *AIEE Conference Paper* presented at the Fall General Meeting, Chicago, Illinois, October 7–11, 1957.
6. Discussion by A. F. Glimn, L. K. Kirchmayer, R. D. Wood of A New Method of Determining Constants for the General Transmission Loss Equations, E. D. Early, R. E. Watson. *AIEE Trans.*, Vol. 74, Part III, 1955, p. 1421.
7. *Higher Mathematics for Engineers and Physicists*, E. S. Sokolnikoff. McGraw-Hill Book Company, Inc., New York, 1941.
8. *Numerical Mathematical Analysis*, J. B. Scarborough. The Johns Hopkins Press, Baltimore, 1955.
9. *Tensor Analysis of Networks*, G. Kron. John Wiley and Sons, New York, 1939.

Problems

Problem 4.1

Recalculate the transmission-loss formula for the system of Figure 4.1 using Muncie as the reference bus and the method outlined in Sections 4.2 and 4.3 assuming the term $P_m H_{mn}(f_m - f_n)P_n$ to be negligible.

Problem 4.2

From Figure 4.2 and the load-current data given in Section 4.2, calculate w' as given by equation 3–91. Compare this value of w' with that obtained in Section 4.2.

Problem 4.3

Using the method of least squares as described in Section 4.9 of Chapter 4, determine the value of the coefficient (a) in the expression

$$y = ax$$

Observed values of y for given values of x are

	x	y (observed)
Case 1	1	0.9
2	2	2.1
3	3	3.2
4	4	3.9

Problem 4.4

Consider the transmission system of Figure 4.20.

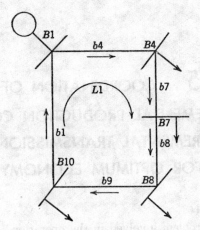

Figure 4.20. Transmission system for problem 4.4.

The branch impedances are given by equation 4–75. Assume $B8$ as the reference bus. Using Method 2 as described in Section 4.10, determine the voltages at all buses with $B8$ grounded and with 1 per unit current entering at bus $B1$.

5 COORDINATION OF INCREMENTAL PRODUCTION COSTS AND INCREMENTAL TRANSMISSION LOSSES FOR OPTIMUM ECONOMY

5.1 INTRODUCTION

An important problem involved in the operation of large integrated power systems is the determination of generation schedules for optimum system economy, including the effects of both incremental production costs and incremental transmission losses. This chapter discusses the use of a loss formula (such as described in Chapters 3 and 4) in coordinating incremental production costs and incremental transmission losses.[1] In this chapter it is intended

1. To present a mathematical analysis of various methods of coordinating incremental production costs and incremental transmission losses.
2. To evaluate the errors introduced in optimum system operation by the assumptions involved in the various methods of coordination.
3. To evaluate the errors introduced in optimum system operation by the assumptions involved in determining a transmission-loss formula.
4. To evaluate the savings obtained in system operation by coordinating incremental production costs and incremental transmission losses.

5.2 COORDINATION OF INCREMENTAL PRODUCTION COSTS AND INCREMENTAL TRANSMISSION LOSSES

In order to combine incremental production costs and incremental transmission losses it is necessary first to express the incremental transmission losses in terms of incremental costs. As shown in Appendix 5, the incremental transmission losses should be charged at a rate equal to the incremental cost of received power.

Three methods of coordinating incremental production costs and incremental transmission losses are discussed:

COORDINATION METHODS FOR OPTIMUM ECONOMY 161

EXACT METHOD INVOLVING NONLINEAR SIMULTANEOUS EQUATIONS

The minimum input in dollars per hour for a given received load is obtained by solution of the following simultaneous equations:

$$\frac{dF_n}{dP_n} + \lambda \frac{\partial P_L}{\partial P_n} = \lambda \qquad (5\text{-}1)$$

or

$$\frac{1}{\lambda}\frac{dF_n}{dP_n} + \frac{\partial P_L}{\partial P_n} = 1 \qquad (5\text{-}2)$$

where

$F_n =$ input to plant n in dollars per hour

$P_n =$ output of plant n in megawatts

$\dfrac{dF_n}{dP_n} =$ incremental production cost of plant n in dollars per mw-hr

$P_L =$ total transmission losses

$\dfrac{\partial P_L}{\partial P_n} =$ incremental transmission loss at plant n in megawatts per megawatt

$\lambda =$ incremental cost of received power in dollars per mw-hr

In general, the incremental transmission loss at plant n may be expressed by

$$\frac{\partial P_L}{\partial P_n} = \sum_m 2B_{mn}P_m$$

where $B_{mn} =$ transmission-loss-formula constants.

The incremental production cost of a given plant over a limited range may be represented by

$$\frac{dF_n}{dP_n} = F_{nn}P_n + f_n$$

$F_{nn} =$ slope of incremental production-cost curve

$f_n =$ intercept of incremental production-cost curve

Then equation 5-1 becomes

$$F_{nn}P_n + f_n + \lambda \sum_m 2B_{mn}P_m = \lambda \qquad (5\text{-}3)$$

Solutions for different total loads are obtained by varying the magnitude of λ.

ECONOMIC OPERATION OF POWER SYSTEMS

APPROXIMATE METHOD INVOLVING LINEAR SIMULTANEOUS EQUATIONS

The incremental transmission loss at plant n in equation 5-1 is charged at λ, the incremental cost of the received power. If the incremental transmission loss at plant n is charged at a constant rate β, the following set of linear simultaneous equations results:

$$\frac{dF_n}{dP_n} + \beta \frac{\partial P_L}{\partial P_n} = \lambda \qquad (5\text{-}4)$$

where β = average incremental cost of received power in dollars per mw-hr.

Equation 5-4 may also be written

$$\frac{1}{\beta}\frac{dF_n}{dP_n} + \frac{\partial P_L}{\partial P_n} = \frac{\lambda}{\beta} = \phi \qquad (5\text{-}5)$$

where

$$\phi = \frac{\lambda}{\beta}$$

This set of linear simultaneous equations corresponds to those described by George, Page, and Ward.[2] Solutions for different total loads are obtained by varying the value of $\phi = \dfrac{\lambda}{\beta}$. Whenever $\dfrac{\lambda}{\beta} = \phi = 1$, an exact solution is obtained corresponding to the solution of the nonlinear simultaneous equations.

PENALTY-FACTOR METHOD

From equation 5-1,

$$\frac{dF_n}{dP_n} + \lambda \frac{\partial P_L}{\partial P_n} = \lambda \qquad (5\text{-}1)$$

Then

$$\frac{dF_n}{dP_n} = \lambda - \lambda \frac{\partial P_L}{\partial P_n} = \lambda \left(1 - \frac{\partial P_L}{\partial P_n}\right)$$

Thus

$$\frac{dF_n}{dP_n}\left[\frac{1}{1 - (\partial P_L/\partial P_n)}\right] = \lambda \qquad (5\text{-}6)$$

$$\frac{dF_n}{dP_n} L_n = \lambda$$

COORDINATION METHODS FOR OPTIMUM ECONOMY

where L_n = penalty factor of plant n

$$= \frac{1}{[1 - (\partial P_L/\partial P_n)]}$$

The solution of equation 5–6 leads to results which, of course, are identical to equation 5–1.

Equation 5–6 may be approximated by

$$\frac{dF_n}{dP_n}\left(1 + \frac{\partial P_L}{\partial P_n}\right) = \lambda \qquad (5\text{-}7)$$

where

$$L_n' = 1 + \frac{\partial P_L}{\partial P_n}$$

= approximate penalty factor of plant n.

Equation 5–7 may be written as

$$\frac{dF_n}{dP_n} + \frac{dF_n}{dP_n}\frac{\partial P_L}{\partial P_n} = \lambda$$

From this equation it is seen that the approximate penalty-factor method [3] corresponds to charging the incremental transmission loss of plant n at a rate corresponding to the incremental production cost of plant n.

PHYSICAL INTERPRETATION OF COORDINATION EQUATIONS

The physical interpretation of the preceding coordination equations 5–1 may be visualized by inspection of Figure 5.1. The incremental

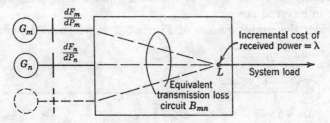

Figure 5.1. Schematic representation of cost relations.

production cost of a given plant n is measured at the plant bus and is denoted by dF_n/dP_n. A given plant n incurs an incremental transmission loss $\partial P_L/\partial P_n$ in supplying the next increment of system load. It is desired that the incremental cost of the power received from each plant be the same at the receiver point L.

For example, suppose that the load increases by an amount ΔP_R. Assume that this load change is first taken up by plant 1 only by increasing the output of plant 1 by ΔP_1. Then the cost of this increment of power at the receiver is given by

$$\lambda = \frac{dF_1}{dP_1} \frac{\Delta P_1}{\Delta P_R}$$

The expression $\Delta P_1 / \Delta P_R$ may be thought of as the reciprocal of the incremental efficiency of the transmission system. The foregoing equation may be rewritten as

$$\lambda = \frac{dF_1}{dP_1} \frac{\Delta P_1}{\Delta P_1 - \Delta P_L}$$

$$= \frac{dF_1}{dP_1} \frac{1}{1 - (\Delta P_L / \Delta P_1)}$$

As ΔP_1 becomes progressively smaller, we have

$$\lambda = \frac{dF_1}{dP_1} \left(\frac{1}{1 - (\partial P_L / \partial P_1)} \right)$$

$$= \frac{dF_1}{dP_1} L_1$$

which is the same form as equation 5–6. In general, the penalty factor of plant n is the ratio of the small change in power at plant n to the small change in received power when only plant n supplies this small change in received power.

Consider the simple system of Figure 5.2 in which we have a single

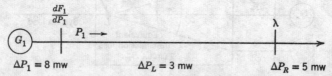

Figure 5.2. Simple one-plant system.

plant supplying a single load. It is desired to know the incremental cost of received power. Let $\Delta P_R = 5$ mw. Assume that to supply this increase in received power it is necessary to increase P_1 by 8 mw. Then

$\Delta P_L = 8$ mw $- 5$ mw $= 3$ mw. If the cost at the plant bus bar is assumed to be three dollars per mw-hr, then the cost at the receiver is given by

$$\lambda = 3 \times \frac{\Delta P_1}{\Delta P_R} = 3 \times \frac{8}{5} = 4.8 \text{ dollars per mw-hr}$$

or

$$\lambda = 3 \times \frac{1}{1 - (\Delta P_L/\Delta P_1)} = 3 \times \frac{1}{1 - (\frac{3}{8})}$$

$$= 3 \times \frac{1}{\frac{5}{8}} = 4.8 \text{ dollars per mw-hr}$$

5.3 APPLICATION OF COORDINATION EQUATIONS TO A TWO-PLANT SYSTEM

To assist in evaluating the various methods described, application is made first to the simple two-plant system shown in Figure 2.29. This system is a simple representation of the 1950 American Gas and Electric System and illustrates the relatively high-cost generation in the Indiana Division, as compared to low-cost generation in the Ohio Division, available for transfer west. These two areas are interconnected by a 132-kv transmission system and are separated by a distance of approximately 250 miles. The maximum practical transfer over the transmission system corresponds to approximately 170 mw.

The approximate incremental production costs for plants 1 and 2 are shown in Figure 2.30 and may be expressed by

$$\frac{dF_1}{dP_1} = \text{incremental production cost of plant 1 in dollars per mw-hr}$$

$$= F_{11}P_1 + f_1$$

$$= 0.002P_1 + 1.7$$

$$\frac{dF_2}{dP_2} = \text{incremental production cost of plant 2 in dollars per mw-hr}$$

$$= F_{22}P_2 + f_2$$

$$= 0.004P_2 + 2.0$$

The values of P_1 and P_2 are to be given in megawatts. The total transmission losses are given by

$$P_L = \text{total transmission losses}$$

$$= B_{11}P_1{}^2 + B_{22}P_2{}^2 + 2B_{12}P_1P_2$$

ECONOMIC OPERATION OF POWER SYSTEMS

The incremental transmission losses at plants 1 and 2 are given by

$$\frac{\partial P_L}{\partial P_1} = \text{incremental transmission loss of plant 1}$$
$$= 2B_{11}P_1 + 2B_{12}P_2$$
$$\frac{\partial P_L}{\partial P_2} = \text{incremental transmission loss of plant 2}$$
$$= 2B_{12}P_1 + 2B_{22}P_2$$

where
$$B_{11} = 0.002$$
$$B_{22} = 0$$
$$B_{12} = 0$$

These constants correspond to a 20-mw loss for a transfer of 100 mw. The exact nonlinear equations for a two-plant system are given by inspection of equation 5-1 as

$$\left.\begin{array}{l} F_{11}P_1 + \lambda(2B_{11}P_1 + 2B_{12}P_2) = \lambda - f_1 \\ F_{22}P_2 + \lambda(2B_{12}P_1 + 2B_{22}P_2) = \lambda - f_2 \end{array}\right\} \quad (5\text{-}8)$$

From equation 5-4 the approximate linear equations may be written

$$\left.\begin{array}{l} F_{11}P_1 + \beta(2B_{11}P_1 + 2B_{12}P_2) = \lambda - f_1 \\ F_{22}P_2 + \beta(2B_{12}P_1 + 2B_{22}P_2) = \lambda - f_2 \end{array}\right\} \quad (5\text{-}9)$$

The application of the approximate penalty-factor method gives the following equations:

$$\left.\begin{array}{l} (F_{11}P_1 + f_1)L_1' = \lambda \\ (F_{22}P_2 + f_2)L_2' = \lambda \end{array}\right\} \quad (5\text{-}10)$$

where
$$L_1' = \text{approximate penalty factor of plant 1}$$
$$= 1 + 2B_{11}P_1 + 2B_{12}P_2$$
$$L_2' = \text{approximate penalty factor of plant 2}$$
$$= 1 + 2B_{12}P_1 + 2B_{22}P_2$$

If the effect of the transmission losses are neglected in the scheduling of generation, then optimum loading is given by

$$\left.\begin{array}{l} F_{11}P_1 + f_1 = \lambda \\ F_{22}P_2 + f_2 = \lambda \end{array}\right\} \quad (5\text{-}11)$$

By choosing appropriate values of λ, various solutions for the foregoing sets of coordination equations were obtained and the resulting generation schedules were plotted, as shown in Figures 5.3 and 5.4. It is to

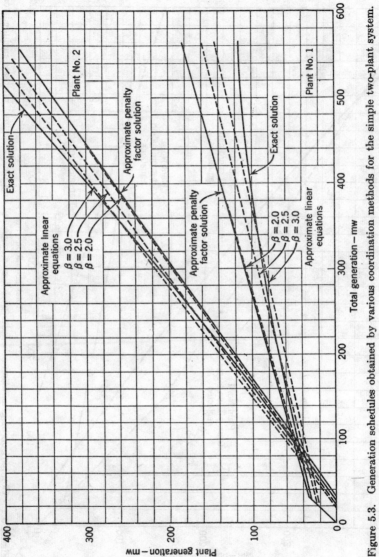

Figure 5.3. Generation schedules obtained by various coordination methods for the simple two-plant system.

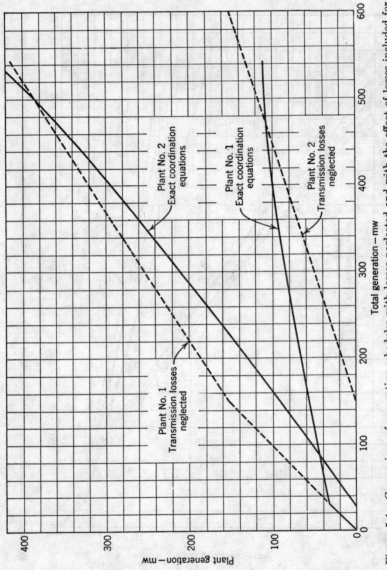

Figure 5.4. Comparison of generation schedules with losses neglected and with the effect of losses included for simple two-plant system.

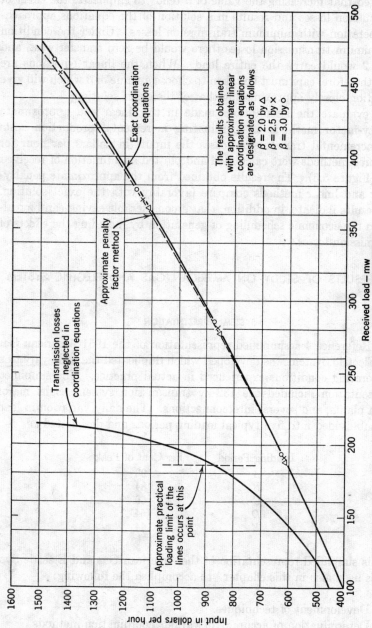

Figure 5.5. Fuel input for the various generation schedules determined for the simple two-plant system.

be noted that increasing the value of β tends to emphasize the effect of transmission losses and results in a solution of the equations approaching operation with minimum transmission losses. Under the condition of minimum transmission losses there would be zero transfer west, and plant 2 would carry the entire load. When the linear equations are used, therefore, care must be taken to choose a value of β which will give a solution closely approximating that of the exact solution.

To evaluate the assumptions made in the linear and approximate penalty-factor methods of coordinating incremental production costs and incremental transmission losses the inputs in dollars per hour for the three methods were calculated and plotted as a function of received load (Figure 5.5). The results obtained from the approximate penalty-factor and linear methods compare favorably with the exact solution. The results indicate, in addition, that a considerable saving can be realized in the economic scheduling of generation by including the effects of transmission losses.

5.4 RESULTS OF STUDY ON AMERICAN GAS AND ELECTRIC SYSTEM

SYSTEM REPRESENTATION

In Reference 4 a simplified representation of the 1951 American Gas and Electric System was developed which duplicated the wide variation of operating conditions experienced in actual practice. The simplified representation included the 132-kv transmission system, eight major steam plants, and several interconnections. The daily variation in load was subdivided into five typical loading periods and designated as

Loading Period	Per Cent of Peak
A	93
B	84
C	69
D	54
E	44

This simplified representation of the American Gas and Electric System is used also in this chapter [1] to accomplish the following:

1. Development of techniques.
2. Determination of accuracy of various coordination methods.
3. Evaluation of savings of a large power system obtained by including the effect of transmission losses in the scheduling of generation.

TABLE 5.1. TRANSMISSION-LOSS-FORMULA COEFFICIENTS ($B_{nm} \times 10^2$)

m	n	D Loading	B Loading	Average
1	1	0.08670	0.07055	0.07863
2	2	0.06145	0.06050	0.06098
3	3	0.11240	0.07090	0.09163
4	4	0.02730	0.02565	0.02646
5	5	0.02390	0.02235	0.02311
6	6	0.03815	0.03630	0.03723
7	7	0.05950	0.06620	0.06285
8	8	0.12140	0.11870	0.12010
1	2	−0.00986	−0.01011	−0.00999
1	3	−0.01697	−0.01107	−0.01402
1	4	−0.00661	−0.00730	−0.00695
1	5	−0.00898	−0.01372	−0.01136
1	6	−0.02064	−0.02089	−0.02076
1	7	−0.02914	−0.02868	−0.02892
1	8	−0.03498	−0.03086	−0.03292
2	3	0.04688	0.04558	0.04624
2	4	0.01299	0.01194	0.01246
2	5	−0.01227	−0.01208	−0.01218
2	6	−0.01954	−0.01665	−0.01810
2	7	−0.02192	−0.01308	−0.01750
2	8	−0.02276	−0.01232	−0.01754
3	4	0.01314	0.01169	0.01242
3	5	−0.01121	−0.01274	−0.01198
3	6	−0.02630	−0.01778	−0.02204
3	7	−0.03595	−0.01500	−0.02530
3	8	−0.04240	−0.01442	−0.02841
4	5	0.00231	0.00127	0.00179
4	6	−0.00754	−0.00660	−0.00707
4	7	−0.01094	−0.00656	−0.00876
4	8	−0.01292	−0.00692	−0.00992
5	6	0.01181	0.01267	0.01224
5	7	0.00681	0.00769	0.00721
5	8	0.00332	0.00423	0.00378
6	7	0.02210	0.02122	0.02166
6	8	0.01732	0.01632	0.01682
7	8	0.05595	0.05940	0.05768

Note. The B loading corresponds approximately to a total load of 84 per cent of peak; the D loading, 54 per cent of peak.

INCREMENTAL TRANSMISSION LOSS AND INCREMENTAL PRODUCTION-COST DATA

In deriving the loss-formula coefficients for the simplified system the following assumptions were made:

1. The equivalent load current at any bus remains a constant complex fraction of the total equivalent load current. The equivalent load current at a bus is defined as the sum of the line-charging, synchronous condenser, and load current at that bus.
2. The generator-bus voltage magnitudes are assumed to remain constant.
3. The ratio of reactive-to-real power of any source is assumed to remain at a fixed value.
4. The generator-bus angles are assumed to remain constant.

The transmission-loss-formula constants (B_{mn}) were calculated from the D and B loading conditions. Differences occur between the transmission-loss-formula coefficients for these two periods because of changes in the load pattern and in the generator voltages, angles, and power factors. The D loading and B loading transmission-loss-formula coefficients and also their average are given in Table 5.1. Except when otherwise specified, the average B_{mn} loss-formula coefficients were used in this study.

Over a given range of plant output the incremental production cost of plant n may be approximated by the following relationship:

$$\text{incremental production cost of plant } n = \frac{dF_n}{dP_n}$$

$$= F_{nn}P_n + f_n$$

The coefficients in Table 5.2 are valid for the range of loading presented in this chapter.

TABLE 5.2

Plant n	F_{nn}	f_n
1	0.00820	1.280
2	0.00440	0.795
3	0.00190	1.809
4	0.00429	0.657
5	0.00222	0.889
6	0.01200	0.300
7	0.02080	0.635
8	0.01270	0.572

COORDINATION METHODS FOR OPTIMUM ECONOMY 173

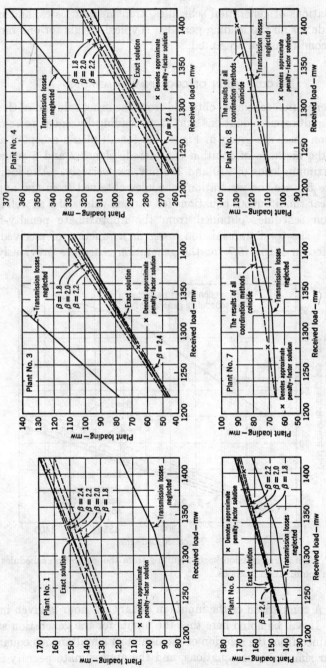

Figure 5.6. Generation schedules obtained by various coordination methods.

Calculations of generation schedules, transmission losses, and inputs were made for all the loading periods. Typical results for one range of loading conditions are presented.

COMPARISON OF COORDINATION METHODS

The generation schedules obtained from the exact solution of the nonlinear simultaneous equations and from loading by equal incremental production costs are given in Figure 5.6 for all plants except 2 and 5. For all the methods of solution considered, plants 2 and 5 remain at their maximum loads of 210 and 310 mw, respectively. Also indicated in Figure 5.6 are the generation schedules obtained from the approximate linear simultaneous equations for various values of β as well as the generation schedules obtained from the approximate penalty-factor method. The total transmission losses as a function of received load are plotted in Figure 5.7 for the generation schedules previously dis-

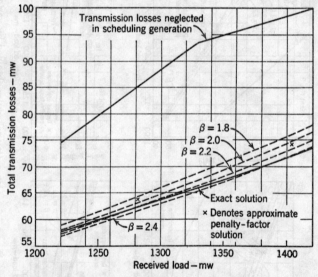

Figure 5.7. Total transmission losses resulting from the generation schedules given by various coordination methods.

cussed. A comparison of the inputs in dollars per hour is given in Figure 5.8. It will be noted here that the inputs for the generation schedules obtained from the solution of the nonlinear simultaneous equations, the linear simultaneous equations, and the approximate penalty-factor method are practically identical.

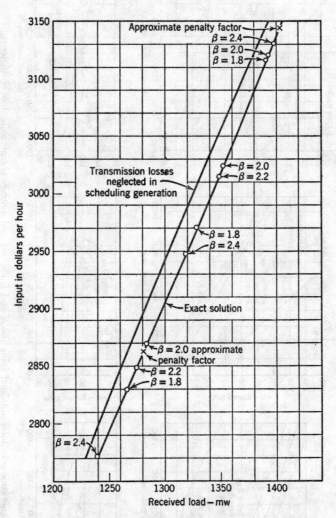

Figure 5.8. Inputs for the various generation schedules.

176   ECONOMIC OPERATION OF POWER SYSTEMS

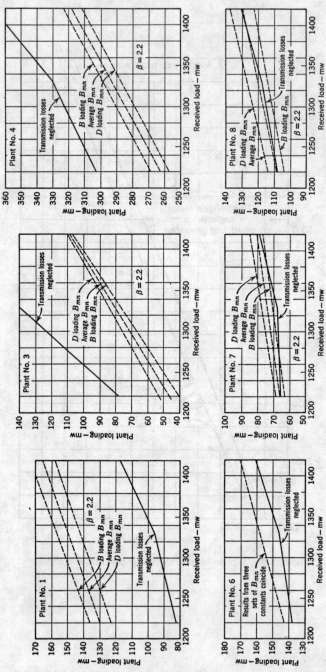

Figure 5.9. Generation schedules obtained using the B and D loading and average transmission loss coefficients.

COORDINATION METHODS FOR OPTIMUM ECONOMY 177

ANALYSIS OF EFFECTS OF VARIATIONS IN LOSS-FORMULA COEFFICIENTS

With $\beta = 2.2$, the effect of the variations in the B_{mn} constants that occur between the B and D loading periods was investigated. The generation schedules obtained from equation 5–4 with the B loading, the D loading, and the average loss-formula coefficients are compared with the loading by equal incremental production costs in Figure 5.9. A corresponding comparison of transmission losses and inputs is given in Figures 5.10 and 5.11. It will be noted from Figures 5.9, 5.10, and 5.11 that

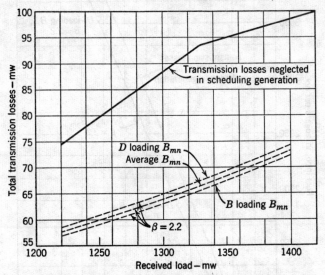

Figure 5.10. Total transmission losses for the schedules obtained using the B and D loading and average transmission loss coefficients in the coordination equations.

the variation of the B_{mn} constants which occurs between the different loading periods causes little change in the economic scheduling of generation.

EVALUATION OF ANNUAL SAVINGS

For the five loading periods the dollars saved per hour by including the effect of the incremental transmission losses over the method of scheduling generation by equal incremental production costs were calculated and plotted as shown in Figure 5.12. Integrated over a year, a saving of approximately 150,000 dollars is to be realized.

178   ECONOMIC OPERATION OF POWER SYSTEMS

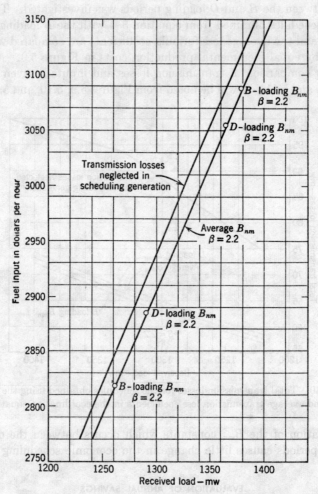

Figure 5.11. Inputs for the various generation schedules.

COORDINATION METHODS FOR OPTIMUM ECONOMY   179

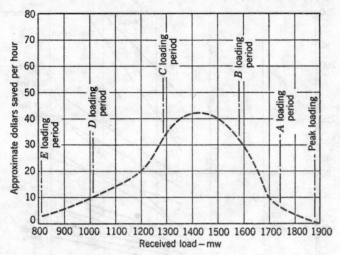

Figure 5.12. Dollars saved in input by including the effect of transmission losses in the economic scheduling of generation.

5.5 USE OF GENERAL LOSS FORMULA

In order to determine any additional savings that may be obtained through the use of a general loss formula of the form

$$P_L = \sum_m \sum_n P_m B_{mn} P_n + \sum_n P_n B_{no} + B_{oo} \qquad (5\text{-}12)$$

over a loss formula of the form

$$P_L = \sum_m \sum_n P_m B_{mn} P_n \qquad (5\text{-}13)$$

a study [5] of the application of these two formulas to the American Gas and Electric System was undertaken. A simplified representation consisting principally of the Ohio and Indiana Divisions of the 1955 American Gas and Electric System was used. The resultant system contained six generating plants, seventeen loads, and approximately 3000 circuit miles of 132-kv and 400 circuit miles of 330-kv transmission. The peak load of this system was approximately 2500 mw. In order to study the actual load variations actual hourly substation loadings for all 132-kv stations in the Ohio and Indiana Divisions were recorded. This load data included readings for a week end and two weekdays for a winter and a summer period. The winter load variation for all loads is shown in Figure 5.13. Substantially the same pattern shown in Figure 5.13 was obtained for the summer load period.

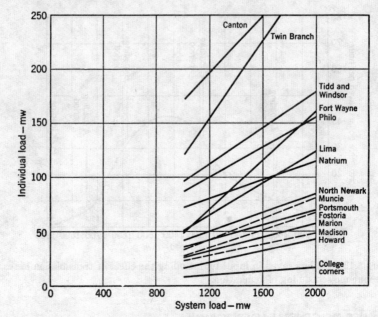

Figure 5.13. Winter load variations.

Two methods of approximating the actual load variation in this system in the derivation of a loss formula were considered. One assumes that all load currents vary together as a constant complex fraction of the total load current and takes the form

$$P_L = \sum_m \sum_n P_m B_{mn} P_n \text{ for total system losses} \quad (5\text{-}13)$$

and

$$\frac{\partial P_L}{\partial P_n} = \sum_m 2 B_{mn} P_m \text{ for incremental losses} \quad (5\text{-}14)$$

These coefficients were calculated as outlined in Section 4.2. The second type of loss formula assumes that the load currents vary linearly between their actual maximums and minimums and takes the form

$$P_L = \sum_m \sum_n P_m B_{mn} P_n + \sum_n P_n B_{no} + B_{oo} \text{ for total system losses} \quad (5\text{-}12)$$

and $\dfrac{\partial P_L}{\partial P_{\dot n}} = \sum_m 2 P_m B_{mn} + B_{no}$ for incremental losses $\quad (5\text{-}15)$

These coefficients were calculated as outlined in Section 4.9.

For the study of this simplified representation of the Ohio and Indiana Divisions of the American Gas and Electric System it was found that the additional savings obtained by use of loss-formula equation 5-12 over those obtained by loss-formula equation 5-13 are marginal.

5.6 CONCLUSIONS

From the typical results presented and similar data from other loading periods, the following conclusions can be drawn concerning the system studied:

1. The operating economy obtained by scheduling generation by the linear simultaneous equations and the approximate penalty-factor method is for all practical purposes identical to that obtained by the solution of the exact nonlinear equations.

2. An average set of loss-formula constants may be applied over the complete daily cycle in the coordination of incremental production costs and incremental transmission losses.

3. The differences in the results obtained by use of D and B loading coefficients indicated in Figures 5.9, 5.10, and 5.11 become even more negligible if the reactive characteristics of the necessary plants are represented by

$$Q_n = Q_{Ln} + s_n P_n$$
$$= \text{plant reactive output}$$

instead of $\quad Q_n = s_n P_n$

where $\quad Q_{Ln} =$ component of plant reactive that is included with the load

4. For a large integrated power system savings of considerable magnitude can be realized when the effect of transmission losses are included in the economic scheduling of generation.

5.7 SUMMARY

Optimum economy is obtained when the incremental cost of received power at the system load is the same from each source. This may be stated mathematically as

$$\frac{dF_n}{dP_n}\left[\frac{1}{1 - (\partial P_L/\partial P_n)}\right] = \lambda \tag{5-6}$$

where $\quad \dfrac{dF_n}{dP_n} =$ incremental production cost of plant n in dollars per mw-hr

$\dfrac{\partial P_L}{\partial P_n} =$ incremental transmission loss at plant n in megawatts per megawatt

$\lambda =$ incremental cost of received power in dollars per mw-hr

The quantity $\partial P_L/\partial P_n$ may be calculated from a transmission-loss formula as

$$\frac{\partial P_L}{\partial P_n} = \sum_m 2P_m B_{mn} + B_{no}$$

The term $[1/1 - (\partial P_L/\partial P_n)]$ is frequently referred to as a penalty factor. In general, the penalty factor of plant n may be thought of as the ratio of the small change in power at plant n to the small change in received power when only plant n supplies this small change in received power. In other words, the penalty factor for plant n is the reciprocal of the incremental efficiency of the transmission network with respect to supplying an increment of system load from plant n.

APPENDIX 5. DETERMINATION OF COORDINATION EQUATIONS

The derivation of these equations follows directly from the method of Lagrangian multipliers described by Courant.[6] Let

$$F_t = \text{total input to system in dollars per hour}$$
$$= \sum_n F_n$$

where F_n = input to plant n in dollars per hour

Let P_L = total transmission losses in megawatts
$$= \sum_m \sum_n P_m B_{mn} P_n$$

where P_n = loading of plant n

B_{mn} = transmission-loss-formula coefficients

It is desired to minimize the total input (F_t) in dollars per hour for a given received load (P_R). Let

$$P_R = \text{given received load}$$

By application of the method of Lagrangian multipliers the equation of constraint is given by

$$\Psi(P_1, P_2, P_3 \cdots P_n) = \sum_n P_n - P_L - P_R = 0 \quad (5\text{--}16)$$

Then minimum input for a given received load is obtained when

$$\frac{\partial \mathcal{F}}{\partial P_n} = 0 \quad (5\text{--}17)$$

where
$$\mathfrak{F} = F_t - \lambda \Psi$$
$$\lambda = \text{Lagrangian type of multiplier} \quad (5\text{--}18)$$

$$\frac{\partial \mathfrak{F}}{\partial P_n} = \frac{\partial F_t}{\partial P_n} - \lambda \frac{\partial \Psi}{\partial P_n} = 0 \quad (5\text{--}19)$$

Then
$$\frac{\partial F_t}{\partial P_n} - \lambda \frac{\partial}{\partial P_n}\left[\sum_n P_n - P_L - P_R\right] = 0$$

$$\frac{\partial F_t}{\partial P_n} - \lambda \left[1 - \frac{\partial P_L}{\partial P_n}\right] = 0$$

$$\frac{\partial F_t}{\partial P_n} + \lambda \frac{\partial P_L}{\partial P_n} = \lambda \quad (5\text{--}20)$$

But
$$\frac{\partial F_t}{\partial P_n} = \frac{\partial \left(\sum_n F_n\right)}{\partial P_n} = \frac{\partial F_n}{\partial P_n} = \frac{dF_n}{dP_n}$$

Then equation 5–20 becomes

$$\frac{dF_n}{dP_n} + \lambda \frac{\partial P_L}{\partial P_n} = \lambda \quad (5\text{--}21)$$

Equation 5–21 is identical to equation 5–1 previously presented.

For the examples presented in this chapter, the flows in the tie lines interconnecting the area studied with foreign areas were considered to be independent of the generation allocation within the area studied. For this assumption, $\partial P_L/\partial P_n$ is given by $\sum_m 2P_m B_{mn}$. If the change in tie flows that occurs when an increment of power is sent from a given generator to the system load is significant, the effect of this change in tie flows is included in the expression for $\partial P_L/\partial P_n$ as

$$\frac{\partial P_L}{\partial P_n} = \sum_m 2P_m B_{mn} + \sum_j \frac{\partial P_j}{\partial P_n} \sum_m 2P_m B_{mj} \quad (5\text{--}22)$$

The quantity $\partial P_j/\partial P_n$ corresponds to the ratio of the change in tie flow j to the change in P_n when an increment of power is sent from P_n to the area load. The magnitude of $\partial P_j/\partial P_n$ may be calculated as described in References 7 and 8.

References

1. Evaluation of Methods of Co-ordinating Incremental Fuel Costs and Incremental Transmission Losses, L. K. Kirchmayer, G. W. Stagg. *AIEE Trans.*, Vol. 71, Part III, 1952, pp. 513–520.

2. Co-ordination of Fuel Cost and Transmission Losses by Use of the Network Analyzer to Determine Plant Loading Schedules, E. E. George, H. W. Page, J. B. Ward. *AIEE Trans.*, Vol. 68, Part II, 1949, pp. 1152–1160.
3. Transmission Losses and Economic Loading of Power Systems, L. K. Kirchmayer, G. H. McDaniel. *General Electric Review*, Schenectady, New York, October 1951.
4. Analysis of Total and Incremental Losses in Transmission Systems, L. K. Kirchmayer, G. W. Stagg. *AIEE Trans.*, Vol. 70, Part I, 1951, pp. 1197–1205.
5. Accuracy Considerations in Economic Dispatching of Power Systems, A. F. Glimn, L. K. Kirchmayer, V. R. Peterson, G. W. Stagg. *AIEE Trans.*, Vol. 75, Part III, 1956, pp. 1125–1137.
6. *Differential and Integral Calculus*, R. Courant. Interscience Publishers, New York, Vol. II, 1936, pp. 188–211.
7. Analysis of Losses in Loop-Interconnected Systems, A. F. Glimn, L. K. Kirchmayer, G. W. Stagg. *AIEE Trans.*, Vol. 72, Part III, 1953, pp. 796–807.
8. Improved Method of Interconnecting Transmission Loss Formulas, A. F. Glimn, L. K. Kirchmayer, J. J. Skiles. *AIEE Trans. Paper* 58–513 presented at Great Lakes District Meeting, East Lansing, Michigan, May 1958.

Problems

Problem 5

Assume that units 1 and 2, for which the data is given in problem 2.1, are located as shown in Figure 5.14.

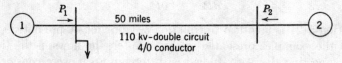

Figure 5.14. Transmission system for problem 5.

Since unit 1 is located at the load center, the loss formula will be of the form

$$P_L = B_{22}P_2^2$$

$$= R_{22}I_2^2$$

The calculation of R_{2-2} is as follows:

Resistance of a 50-mile, double-circuit 4/0 conductor $= 50(0.5)/2 = 12.5$ ohms. For a base of 100 mva and 110 kv,

$$\text{base ohms} = 121 \text{ ohms}$$

Then per unit $\qquad R_{22} = 12.5/121 = 0.103$

Assume $\qquad V_2 = 1$ p.u.

$$s_2 = +0.4 \text{ p.u.}$$

Then $\qquad B_{22} = 1/V_2^2(1 + s_2^2)R_{22} = (1.16)(0.103) = 0.12$

Note that B_{22} is given with respect to a base of 100 mva.

With the data of Problem 2.1 and a knowledge of B_{22}, determine the incremental cost of received power from unit 2, using equation 5–6. Plot this received cost and also the incremental cost of unit 1 as a function of each unit output. These curves represent the incremental cost of received power from units 2 and 1, respectively. From this figure determine the economic allocation of generation for a range of total received load from 50 to 185 mw. Plot P_1 and P_2 as a function of $P_1 + P_2$. For your information compare this generation schedule with that of problem 2.2.

6 PRACTICAL CALCULATION, EVALUATION, AND APPLICATION OF ECONOMIC SCHEDULING OF GENERATION

6.1 GENERAL SUMMARY OF METHOD

This section is intended to present a chronological outline of the required data and calculating procedure involved in the analysis of the economic scheduling of system generation.

A. Preparation of Data
1. Electrical-System Data
   (a) Impedance diagram of transmission and subtransmission facilities whose losses are dependent upon the manner in which generation is scheduled.
   (b) Daily load cycles for typical week's operation.
   (c) Load-duration curve for period of operation to be considered. This curve is required in the determination of the annual differences in production costs resulting from different scheduling methods.
   (d) Selection of base-case loading period and tabulation of loads, voltages, and probable generation schedules and interconnection flows for base case.
2. Plant Data
   (a) Thermal characteristics of units and in particular the incremental fuel-rate data on all units.
   (b) Cost of fuel at various plants in cents per million Btu.
   (c) Determination of straight-line equations of incremental production costs of various units.

B. Determination of Transmission-Loss Formula
1. Resistance measurements on open-circuited transmission network.
2. Base-case load-flow data.

CALCULATION OF GENERATION SYSTEM SCHEDULES

3. Transcribing of these data to punched cards if data for steps 1 or 2 are taken from network analyzer.
4. Calculation of loss-formula coefficients on digital calculator.

C. Evaluation of Savings to be Obtained by Considering Transmission Losses in the Scheduling of Generation
   1. Determine generation schedules by equal incremental production costs.
   2. Determine generation schedules by coordination equations.
   3. Determine cost of received power for each schedule.
   4. Determine difference in cost of received power.
   5. Through use of load-duration curve, determine annual savings.

D. Practical Application of Coordination Equations to Power-System Operation
   1. Precalculated generation schedules.
   2. Use of special computers built particularly for use of load dispatcher to calculate schedules as need arises.
   3. Economic automation system which automatically and simultaneously maintains economic allocation of generation, system frequency, and net interchange.

The material considered in points A and B of the foregoing outline has been discussed in previous chapters.

6.2 DETERMINATION OF IMPORTANCE OF TRANSMISSION-LOSS CONSIDERATIONS

It is prudent to evaluate the savings involved when scheduling generation with the effect of incremental transmission losses included for typical operating conditions and typical fuel-cost data. It is suggested that this step be taken before undertaking point D of the outline in order to determine the amount of effort and dollars it is worthwhile to expend for part D.

6.3 SOLUTION OF COORDINATION EQUATIONS

The determination of generation schedules by equal incremental production costs is discussed in Chapter 2. The theory involved in determining the coordination equations to include the effect of both incremental production costs and incremental transmission losses is covered in Chapter 5. Various methods of solving these coordination equations are discussed here. It is necessary to solve these coordination equations in order to undertake step C and also to precalculate generation schedules for various operating conditions, if desired, in step D.

The coordination equations may be solved by analogue or digital methods. We shall first discuss those suitable for analogue computers, such as the network analyzer and the differential analyzer.

6.4 ANALOGUE METHODS

The use of the network analyzer as a mesh-circuit analogue is described in detail in the paper, "Coordination of Fuel Cost and Transmission Loss by Use of the Network Analyzer to Determine Plant Loading Schedules,"[1] by E. E. George, H. W. Page, and J. B. Ward.

From Chapter 5 we may write the following coordination equations for a two-plant system:

$$F_{11}P_1 + \lambda(2B_{11}P_1 + 2B_{12}P_2) = -(f_1 - \lambda)$$
$$F_{22}P_2 + \lambda(2B_{22}P_2 + 2B_{12}P_1) = -(f_2 - \lambda)$$

Regrouping coefficients, we obtain

$$P_1(F_{11} + \lambda 2B_{11}) + P_2(\lambda 2B_{12}) = -(f_1 - \lambda)$$
$$P_1(\lambda 2B_{12}) + P_2(F_{22} + \lambda 2B_{22}) = -(f_2 - \lambda)$$

Defining
$$A_{11} = F_{11} + \lambda 2B_{11}$$
$$A_{12} = \lambda 2B_{12}$$
$$A_{21} = \lambda 2B_{12} \qquad (6\text{--}1)$$
$$A_{22} = F_{22} + \lambda 2B_{22}$$
$$C_1 = -(f_1 - \lambda)$$
$$C_2 = -(f_2 - \lambda)$$

we obtain
$$A_{11}P_1 + A_{12}P_2 = C_1 \qquad (6\text{--}2)$$
$$A_{21}P_1 + A_{22}P_2 = C_2$$

The use of a network-analyzer mesh-circuit analogue is illustrated in Figure 6.1. The P_n are considered to be currents; the A_{mn}, resistances; and the C_n, voltages.

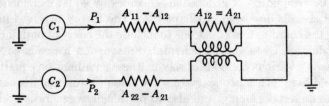

Figure 6.1. Mesh-circuit analogue of coordination equations.

CALCULATION OF GENERATION SYSTEM SCHEDULES

This method has certain practical limitations which restrict its usefulness:

1. Errors introduced by mutual transformers in representing mutual coefficients.
2. Limitation in number of mutual transformers. The use of conductive coupling may eliminate the need for some mutual transformers, but obtaining the proper combination of mutual transformers and conductive coupling is very time consuming.
3. Difficulties involved in representing negative mutual coefficients.
4. Difficulty in maintaining constant current output of a network-analyzer generator when the particular plant in question is at maximum or minimum.
5. Time involved in plugging the analyzer.

If a nodal-circuit analogue is used, the P_n are considered to be voltages; the A_{mn}, inductive or capacitive reactances; and the C_n, currents. The corresponding circuit is given in Figure 6.2. In this analogy the

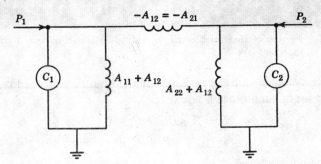

Figure 6.2. Nodal-circuit analogue of coordination equations.

difficulties discussed under points 1, 2, and 3 are usually eliminated. However, an error is introduced because of the resistance in the reactance units. Also, the driving function requires the setting of constant currents with its attendant difficulties, since the network-analyzer generators are constant voltage devices.

The use of an electronic differential analyzer such as that shown in Figure 6.3 is a better analogue method of solving the coordination equations than the a-c network analyzer. Figure 6.3 pictures part of the Reeve's electronic differential-analyzer installation of the General Electric Analytical Engineering Section. The electronic differential analyzer is composed of amplifiers (for adding, inverting, and integrating) and of resistance and capacitance units. In this computer the analogy is be-

Figure 6.3. Portion of General Electric Analytical Engineering Section Electronic Differential Analyzer.

tween d-c voltages and the variables of the system under study. Consider the set of simultaneous equations

$$A_{mn}P_m = C_n \qquad (6\text{-}3)$$

or
$$\boldsymbol{AP} = \boldsymbol{C} \qquad (6\text{-}4)$$

Rewriting equation 6–3,
$$A_{mn}P_n - C_n = 0 \qquad (6\text{-}5)$$

An auxiliary set of equations with the same steady-state solution is given by
$$\frac{d}{dt}P_n = A_{mn}P_m - C_n \qquad (6\text{-}6)$$

Thus, for $m, n = 1, 2$ we have

$$\frac{d}{dt}P_1 = A_{11}P_1 + A_{12}P_2 - C_1 \qquad (6\text{-}7)$$

$$\frac{d}{dt}P_2 = A_{21}P_1 + A_{22}P_2 - C_2$$

The method of solution of these equations is given in Figure 6.4.

The coordination equations may also be solved in another manner by using electronic differential-analyzer elements, as indicated in Figure 6.5.

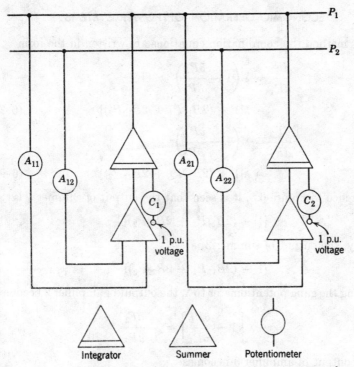

Figure 6.4. Electronic differential analyzer coordination equation analogue.

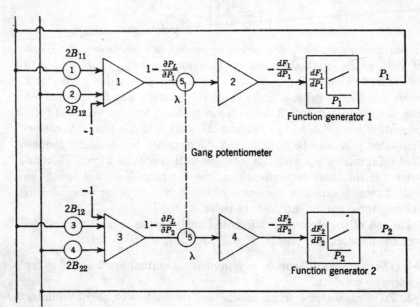

Figure 6.5. Alternative electronic differential analyzer solution of coordination equations.

For this method the coordination equations are written in the form

$$\frac{dF_1}{dP_1} = \lambda\left(1 - \frac{\partial P_L}{\partial P_1}\right)$$

$$= \lambda[1 - (2B_{11}P_1 + 2B_{12}P_2)] \qquad (6\text{-}8)$$

$$\frac{dF_2}{dP_2} = \lambda\left(1 - \frac{\partial P_L}{\partial P_2}\right)$$

$$= \lambda[1 - (2B_{12}P_1 + 2B_{22}P_2)] \qquad (6\text{-}9)$$

By reference to Figure 6.5, it is seen that the output of summer 1 is

$$[1 - (2B_{11}P_1 + 2B_{12}P_2)]$$

Similarly, the output of summer 3 is

$$[1 - (2B_{12}P_1 + 2B_{22}P_2)]$$

By setting the gang potentiometer to λ, the output of amplifier 2 becomes

$$-\lambda\left[1 - \frac{\partial P_L}{\partial P_1}\right] = -\frac{dF_1}{dP_1}$$

and the output of amplifier 3 becomes

$$-\lambda\left[1 - \frac{\partial P_L}{\partial P_2}\right] = -\frac{dF_2}{dP_2}$$

In function generator 1, dF_1/dP_1 is presented as a function of P_1. With dF_1/dP_1 as the input, the output of function generator 1 corresponds to the desired value of P_1. Similarly, the output of function generator 2 yields the desired value of P_2. These values of P_1 and P_2 are then fed into potentiometers 1, 2, 3, and 4, as indicated. Various values of total generation are obtained by changing the setting of the λ potentiometer. The value of λ may be automatically determined by matching the desired total generation with the total generation obtained from the computer. If the total generation from the computer does not equal the desired total generation, the value of λ is driven by a servomechanism to bring these two quantities into balance.

The use of the electronic differential analyzer has several advantages over the network analyzer.

1. There are no difficulties in representing mutual coefficients, either positive or negative.

2. Plant maximums or minimums are automatically held by limiter circuits.

CALCULATION OF GENERATION SYSTEM SCHEDULES

An additional analogue method involves the use of analogue simultaneous equation-solvers. A series of approximations is required for convergence on the final solution. Additional time is required for the solutions, as compared to the methods described earlier, particularly if scaling of the variables is required.

6.5 DIGITAL METHODS

A number of digital methods which may be undertaken manually or by means of digital computers are discussed.

1. Determinants (see your high-school algebra text).
2. Matrix Inversion. If it is assumed that our original set of equations is given by

$$AP = C \qquad (6\text{-}4)$$

then
$$A^{-1}AP = A^{-1}C \qquad (6\text{-}10)$$

$$P = A^{-1}C$$

where
$$A^{-1} = \text{inverse of } A$$

The following method of obtaining an inverse is taken from the book *Tensor Analysis of Networks* by G. Kron,[2] page 29.

(a) Interchange rows and columns.
(b) Replace each element by its minor. The minor of any given element is obtained by striking out the row and column in which the element lies and then calculating the determinant of the remaining matrix.
(c) Multiply every other minor with minus one, starting with plus one in the upper left-hand corner, as shown in the following scheme.

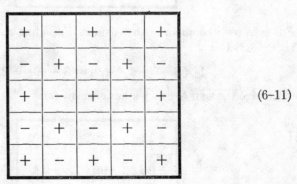

(6-11)

(d) Divide each resulting element by the determinant of the whole original matrix.

For example, assume that the A matrix is of the form

$$A = \begin{vmatrix} 2 & 0.1 & 0 \\ 0.1 & 1 & -0.2 \\ 0 & -0.2 & 3 \end{vmatrix} \qquad (6\text{-}12)$$

By interchanging rows and columns (a), we obtain

$$\begin{vmatrix} 2 & 0.1 & 0 \\ 0.1 & 1 & -0.2 \\ 0 & -0.2 & 3 \end{vmatrix} \qquad (6\text{-}13)$$

(Since the coordination equations result in a symmetrical A matrix, this step may be skipped when working with the coordination equations.)

Replace each element by its minor (b):

$$\begin{vmatrix} +2.96 & +0.3 & -0.02 \\ +0.3 & +6 & -0.4 \\ -0.02 & -0.4 & +1.99 \end{vmatrix} \qquad (6\text{-}14)$$

For example, the minor of the element in the first row and first column is calculated as

$$1 \times 3 - (-.2)(-.2) = 3 - .04 = 2.96$$

Change the sign of every other element (c):

$$\begin{vmatrix} +2.96 & -0.3 & -0.02 \\ -0.3 & +6 & +0.4 \\ -0.02 & +0.4 & +1.99 \end{vmatrix} \qquad (6\text{-}15)$$

The determinant of the A matrix is given by (d)

$$2(2.96) - 0.1(0.3) + 0(-0.02) = 5.89$$

Dividing each element of 6-15 by 5.89, we obtain the matrix

$$A^{-1} = \begin{array}{|c|c|c|} \hline +0.5025 & -0.0509 & -0.0034 \\ \hline -0.0509 & +1.0187 & +0.0679 \\ \hline -0.0034 & +0.0679 & +0.3379 \\ \hline \end{array} \quad (6\text{-}16)$$

As indicated in equation 3-21, this inverse may be checked by the fact that

$$AA^{-1} = I \quad (6\text{-}17)$$

$$\begin{array}{|c|c|c|} \hline 2 & 0.1 & 0 \\ \hline 0.1 & 1 & -0.2 \\ \hline 0 & -0.2 & 3 \\ \hline \end{array} \quad \begin{array}{|c|c|c|} \hline +0.5025 & -0.0509 & -0.0034 \\ \hline -0.0509 & +1.0187 & +0.0679 \\ \hline -0.0034 & +0.0679 & +0.3379 \\ \hline \end{array} = \begin{array}{|c|c|c|} \hline 1 & 0 & 0 \\ \hline 0 & 1 & 0 \\ \hline 0 & 0 & 1 \\ \hline \end{array}$$

3. Starring or Pivotal Condensation. Assume that the original set of equations is given by

$$AP = C \quad (6\text{-}4)$$

Form the composite matrix:

$$\begin{array}{|c|c|} \hline A & C \\ \hline -I & 0 \\ \hline \end{array} \quad (6\text{-}18)$$

The equations are solved by an elimination method which is designated as starring or pivotal condensation. Specifically, a given diagonal term A_{kk} of the A matrix is chosen to be the starred term and the operation

$$A_{ij}{}^* = A_{ij} - \frac{A_{kj}A_{ik}}{A_{kk}}$$

is performed in the composite matrix, converting it to a (*) system or reduced matrix from which the k row and k column have been eliminated. This process is repeated until the only remaining nonzero terms are those of the original zero matrix in the lower right-hand corner of the original composite matrix. When the reduction has been completely carried out these terms in the lower right-hand matrix will have become the P matrix which was desired.

The above method may be easily extended to the determination of A^{-1}.

The disadvantage of the digital methods discussed so far results from the fact that whenever a plant reaches maximum or minimum it is necessary to solve a new set of equations in which the plant at maximum or minimum has been eliminated as a variable. Also, when a given plant loading is of such a value that it is necessary to use a different straight-line approximation of the incremental-production-cost data it is necessary to solve a new set of equations. There are difficulties sometimes in getting initial solutions so that the proper plants are at maximum or minimum and so that the plants are on the correct straight-line approximation of the incremental production-cost data.

4. **Iterative Procedure.**[3] The iterative procedure involves a method of successive approximations which rapidly converge to the correct solution. From equation 5-1 the exact coordination equations are given by

$$\frac{dF_n}{dP_n} + \lambda \frac{\partial P_L}{\partial P_n} = \lambda \qquad (6\text{-}19)$$

$$F_{nn}P_n + f_n + \lambda \sum_m 2B_{mn}P_m = \lambda \qquad (6\text{-}20)$$

Collecting all coefficients of P_n, we obtain

$$P_n(F_{nn} + \lambda 2B_{nn}) = -\lambda \left(\sum_{m \neq n} 2B_{mn}P_m \right) - f_n + \lambda \qquad (6\text{-}21)$$

Solving for P_n, we obtain

$$P_n = \frac{1 - \dfrac{f_n}{\lambda} - \sum_{m \neq n} 2B_{mn}P_m}{\dfrac{F_{nn}}{\lambda} + 2B_{nn}} \qquad (6\text{-}22)$$

CALCULATION OF GENERATION SYSTEM SCHEDULES

The number of required iterations in general is quite small, since the diagonal terms are generally much larger than the off-diagonal terms.

This iterative procedure is illustrated for a simple two-plant system. Assume that the loss-formula coefficients in $1/MW$ units are given by

m	n	B_{mn}
1	1	+0.001
1	2	−0.0005
2	2	+0.0024

Also assume that

$$\frac{dF_1}{dP_1} = \text{incremental production cost of plant 1}$$
$$= F_{11}P_1 + f_1$$
$$\frac{dF_2}{dP_2} = \text{incremental production cost of plant 2}$$
$$= F_{22}P_2 + f_2$$

where

$$F_{11} = +0.01 \qquad f_1 = 2.0$$
$$F_{22} = +0.01 \qquad f_2 = 1.5$$

Substituting the above numbers into equation 6–22,

$$P_1 = \frac{1.0 - (2.0/\lambda) + 0.001 P_2}{0.01/\lambda + 0.002}$$

$$P_2 = \frac{1.0 - (1.5/\lambda) + 0.001 P_1}{0.01/\lambda + 0.0048}$$

To find a point in the generation schedule we choose a λ, say 2.5, and iterate until the P_n have converged to sufficient accuracy. The calculating form is then

$$P_1 = \frac{1.0 - 0.80 + 0.001 P_2}{0.004 + 0.002} = \frac{0.2 + 0.001 P_2}{0.006}$$

$$P_2 = \frac{1.0 - 0.6 + 0.001 P_1}{0.004 + 0.0048} = \frac{0.4 + 0.001 P_1}{0.0088}$$

The calculations are started by first assuming all $P_n = 0$; and as new values are calculated they are used immediately as follows:

Iteration No.	Calculation	
1	$P_1 = \dfrac{0.2 + 0}{0.006}$	$= 33.3$
	$P_2 = \dfrac{0.4 + 0.0333}{0.0088}$	$= 49.2$
2	$P_1 = \dfrac{0.2 + 0.0492}{0.006}$	$= 41.5$
	$P_2 = \dfrac{0.4 + 0.0415}{0.0088}$	$= 50.2$
3	$P_1 = \dfrac{0.2 + 0.0502}{0.006}$	$= 41.7$
	$P_2 = \dfrac{0.4 + 0.0417}{0.0088}$	$= 50.2$
4	$P_1 = \dfrac{0.2 + 0.0502}{0.006}$	$= 41.7$
	$P_2 = \dfrac{0.4 + 0.0417}{0.0088}$	$= 50.2$

The iterative solution is completed after four iterations, since the results of the fourth iteration agree with the preceding results to a sufficient degree of accuracy. The number of required iterations is small, even for a much larger system, as the terms which are independent of the other plants are the dominant ones.

If a knowledge of system transmission losses is desired, they may be evaluated by a total loss formula

$$P_L = P_m B_{mn} P_n = B_{11} P_1^2 + 2 B_{12} P_1 P_2 + B_{22} P_2^2$$

$$= 0.0010(41.7)^2 - (2)(0.0005)(41.7)(50.2)$$

$$+ 0.0024(50.2)^2 = 5.7 \text{ mw}$$

The total generation is obtained by

$$P_T = \sum_n P_n = 41.7 + 50.2 = 91.9 \text{ mw}$$

The received power is calculated

$$P_R = P_T - P_L = 91.9 - 5.7 = 86.2 \text{ mw}$$

CALCULATION OF GENERATION SYSTEM SCHEDULES 199

Thus, one entry in a table used for scheduling is

P_1	P_2	Total Generation	Loss	Received Load	λ
41.7	50.2	91.9	5.7	86.2	2.5

This type of problem is well suited to the use of a digital computer; and generation schedules may be economically calculated by this iterative method either with small- or large-scale, general-purpose digital computers. One computer toward the small end of the spectrum which has been used for a substantial amount of scheduling is the card-programmed calculator (CPC). As explained in Chapter 4, the control unit of the CPC senses holes punched in cards, which are the basis of operation sequencing. Thus a deck of these cards is a program of operations, as the name of the device implies.

The larger scale computer operates in a similar manner, but the numbers which the control unit senses may be stored in its more extensive memory, permitting more rapid access to these instructions and thus a higher effective operating speed. Many such machines, with variations in the techniques of storage in memory and data-handling equipment, are in existence. The use of either size of equipment may be justified for scheduling, depending on the circumstances.

It is desirable that the computer programmer see the logical sequence of operations in a problem represented graphically. This is usually called a flow diagram, and one form is illustrated in Figure 6.6. The sequence of basic operations is established by following the arrows through operation boxes and taking the correct branch when an operation box has two output paths. The operations pertaining to data handling were omitted, since they are not the same for all computers. Thus the form shown permits the use of a wide range of general-purpose digital computers for the scheduling of generation.

This digital-computer method also provides the automatic selection of the appropriate incremental-cost characteristics. As previously discussed, the representation of the incremental-cost characteristic is achieved by breaking the incremental-cost curves into several sections which may be accurately represented by straight-line segments. The computer can now select the side of the break point on which the previously iterated generation of that plant occurred and thus find the correct slope and intercept to use in the current iteration. Also, the plants have physical maximum capacities and some practical minimum-load points. These limitations must be imposed, and a digital computer can do this very well by testing the generation calculated. If the re-

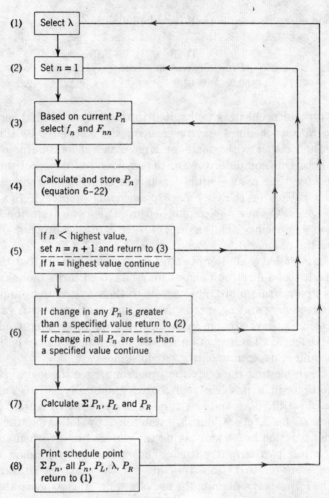

Figure 6.6. Digital computer flow diagram for solution of coordination equations.

sulting value is outside the limits, the computer substitutes the appropriate limiting value.

In using the computer for scheduling one may arbitrarily select a set of values for λ and proceed to calculate a schedule point for each value of λ. This mode of operation is quite efficient. For example, for one given value of λ, the time on the IBM 650 computer to calculate the economic allocation of generation, total generation, losses, and received load for a ten-plant system would be approximately 10 minutes.

In other situations it may be convenient to calculate the λ which causes

CALCULATION OF GENERATION SYSTEM SCHEDULES   201

1. The net generation of a given plant to be a specified value.
2. The total generation to be a specified value.
3. The received power to be a specified value.

In any of these forms a new value of λ is calculated for each iteration in addition to new generation amounts, and all quantities converge to constant values.

It is desirable in defining a schedule with a minimum number of points to obtain solutions corresponding to those outputs at which individual plants encounter their maximum, break, and minimum generations. In this case the defining equation is

$$\lambda = \frac{F_{nn}P_n + f_n}{1 - 2\sum_m B_{mn}P_m}$$

where P_n is the load on plant n that is specified rather than calculated.

In scheduling on a specified total generation basis the λ are calculated by

$$\lambda^i = \lambda^{i-1} + (P_T{}^d - P_T{}^{i-1})\left(\frac{\lambda^{i-1} - \lambda^{i-2}}{P_T{}^{i-1} - P_T{}^{i-2}}\right) \qquad (6\text{--}23)$$

where superscript i indicates the iteration being started
  $i - 1$ indicates the iteration just completed
  $i - 2$ indicates the preceding iteration
  P_T = total generation ($\sum P_n$)
  $P_T{}^d$ = desired total generation

Scheduling by this scheme is initiated by selecting two values of λ. Iteration is carried out to convergence for both these λ values to establish the $P_T{}^{i-1}$ and $P_T{}^{i-2}$ required in equation 6–23.

Scheduling on a specified received load may be achieved by substituting P_R for P_T in equation 6–23. This implies a substantial increase in calculation, as total losses must be calculated for each iteration. Since most utilities schedule on the basis of total generation rather than received power, this form would not usually be desired except in making economic comparisons.

The iterative digital method described has been found to be an extremely valuable tool in obtaining more economical operation of power systems. This method of solution of coordination equations offers several distinct advantages over previous methods:

1. The flexibility of this method allows a generation schedule point to be obtained for a given value of incremental cost of received power,

or a specific value of total load, a specific value of total generation, or a specific load on a given plant. Thus a total schedule may be obtained with a minimum number of solutions.

2. Once the loss-formula coefficients are developed, the time required to transcribe the data for use in the computer is very small. Since the program of calculation is general, a single routine is maintained in the library which will permit scheduling of any size system in a most efficient manner.

3. For convenience in comparing schedules the incremental cost of received power, total transmission losses, and received load may be calculated and tabulated along with the allocation and summation of generation. If desired, the fuel input to each plant, as well as the total fuel input, may also be calculated.

4. The described method may be applied to all general-purpose digital computers.

6.6 PROCEDURE FOR EVALUATING ANNUAL SAVINGS

The following general procedure for evaluating the annual savings incurred by including transmission-loss considerations in the scheduling of generation is suggested:

1. Determine the generation schedule for operation at equal incremental production costs, neglecting the effect of transmission losses.

2. Determine the generation schedule, including the effects of both incremental production costs and incremental transmission losses.

3. Calculate cost of received power by the following method:

   (a) Determine P_L by summing the i^2R losses, line-by-line, or from a transmission-loss formula.

   Thus
   $$P_L = \sum i_k^2 R_k \tag{6-24}$$
   or
   $$P_L = P_m B_{mn} P_n \tag{6-25}$$

   (b) Then the received power P_R is given by
   $$P_R = \sum P_n - P_L \tag{6-26}$$

   (c) The fuel input to a given plant $n(F_n)$ in dollars per hour may be determined by reference to plant heat rate or input-output data. An alternative procedure is to integrate the incremental fuel-cost data. Thus, if
   $$\frac{dF_1}{dP_1} = F_{11} P_1 + f_1$$
   $$F_1 = \frac{F_{11} P_1^2}{2} + f_1 P_1 \tag{6-27}$$

CALCULATION OF GENERATION SYSTEM SCHEDULES 203

This expression for F_1 does not include the fuel input incurred for zero output. If the same units are in operation for both schedules, the no-load, fuel-input intercept is not required, since the input for zero output disappears when subtracting the fuel inputs for such schedules. The total fuel input (F_t) is, of course, given by

$$F_t = \sum F_n \qquad (6\text{-}28)$$

(d) The cost of the received power in dollars per mw-hr is given by

$$F_t/P_R$$

4. Plot F_t/P_R vs. P_R of schedules 1 and 2 (Figure 6.7).

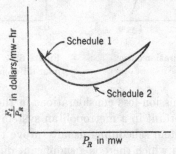

Figure 6.7. Plot of dollars per mw-hr of received load for two schedules.

5. Plot a curve of difference between the two curves of point 4 vs. P_R (Figure 6.8).

6. From 5, plot Δ dollars per hour vs. P_R where

$$\Delta \text{ dollars per hour} = \Delta \text{ dollars per mw-hr} \times P_R$$

(Figure 6.9).

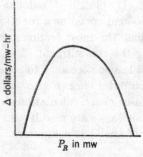

Figure 6.8. Difference in dollars per mw-hr of received load for two schedules.

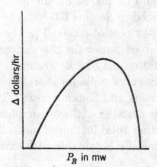

Figure 6.9. Difference in dollars per hr for two schedules.

7. With a load-duration curve (as indicated in Figure 6.10) plot Δ dollars per hour vs. hours. The integral of this curve is dollars saved (Figure 6.11).

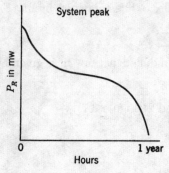

Figure 6.10. Load-duration curve.

Figure 6.11. Plot of dollars per hour difference vs. time.

In general, transmission-loss considerations in the scheduling of generation are not important in a metropolitan system. Transmission-loss considerations in system scheduling usually prove to be significant in a widespread system in which there is a significant difference in the incremental production costs between various parts of the power system. Typical annual savings obtained for widespread systems are fifty thousand dollars per 1000 mw of installed capacity.

6.7 DETERMINATION OF UNITS TO BE PLACED IN OPERATION

In order to determine which combination of units can supply the load most economically and, consequently, the economic order in which units should be placed on the line economic comparisons similar to that described in Section 6.6 must be undertaken. By plotting the dollar-per-mw-hr cost of received power (including no-load costs) as a function of received power for the various combinations, the most economic combination may be determined. In studying the commitment of and/or removal of units for short periods of time it is also necessary to consider starting and banking costs. The digital computer offers a great advantage over the analogue computer for undertaking such calculations, since neither total fuel input nor total transmission losses are readily and economically obtainable with existing designs of analogue computers.

CALCULATION OF GENERATION SYSTEM SCHEDULES 205

6.8 PRACTICAL APPLICATION OF COORDINATION EQUATIONS TO POWER-SYSTEM OPERATION

PRECALCULATED SCHEDULES

The simultaneous coordination equations may be solved on various general-purpose computers, previously described in this chapter, to precalculate tables or charts of generation schedules for direct use by the load dispatcher. It is, of course, necessary to calculate groups of schedules corresponding to various planned station outages. The use of precalculated schedules has been found practical for systems for which interconnection transactions are not frequent and for which the fuel costs across the system vary together.

For some systems it has been noted that transmission-loss considerations may affect the scheduling of only two plants, with the other plants hardly changed. From studies of typical operating conditions an operating guide may be developed to modify the loading of the two plants in question.

USE OF SPECIALIZED COMPUTER

In large integrated systems in which the costs of fuel across the system vary independently of each other and in which there are frequent interconnection transactions it is difficult to preconceive all the various combinations of operating conditions that may occur. Also, the required number of precalculated schedules to cover all these conditions might very well be prohibitive. In such cases a specialized computer designed particularly for the use of the load dispatcher may be economically justified. Several such specialized computers have been developed.[4,5,6,7,8,9,10]

PENALTY-FACTOR COMPUTER [4,5,6]

An analogue computer which calculates the previously described penalty factors for all plants and interconnections has been developed cooperatively by the American Gas and Electric Service Corporation and the General Electric Company. This penalty-factor computer, which was installed in the Central Production and Coordination Office of the American Gas and Electric Service Corporation early in 1955, operates in conjunction with an incremental-fuel-cost slide rule, such as that shown in Figure 6.12.

As explained in Chapter 2, the incremental-fuel-cost slide rule is used to compare the incremental production costs of the various units in the system. This slide rule consists of a logarithmically calibrated incremental-production-cost scale, movable strips for each generating unit, movable fuel-cost-adjustment scales, and a penalty-factor scale. Each movable strip for the generating units indicates a relation between the incremental production cost of the unit and output of the unit.

Figure 6.12. American Gas and Electric Service Corporation incremental-fuel-cost slide rule.

The application of the penalty-factor computer shown in Figure 6.13 by the American Gas and Electric Service Corporation results in savings of over two hundred thousand dollars a year compared to operating the system with transmission losses neglected.

This penalty-factor computer, as shown schematically in Figure 6.14, consists essentially of adjustable power units, representing the various plant loadings and interconnection flows, a loss-formula network, and a central penalty-factor and incremental-transmission-loss instrument. After the power settings are made, the time for computation is negligi-

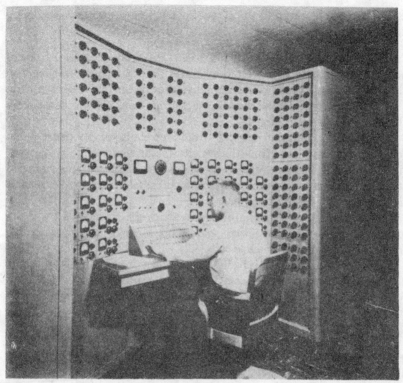

Figure 6.13. Penalty-factor computer installation, Central Production and Coordination Office of American Gas and Electric Service Corporation.

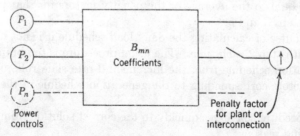

Figure 6.14. Schematic representation of penalty-factor computer.

ble and is little more than the time required to read the penalty factor from an instrument. The penalty factors obtained from the computer are used to adjust the relative positions of the plant strips on the slide rule, and the desired generation schedule is read off the movable generator strips.

The method involved in the actual hour-by-hour economic scheduling of generation is as follows:

1. Set the B_{mn} coefficients into the penalty-factor computer. It is not necessary to change these B_{mn} coefficients, except for major changes in the transmission system which normally occur because of system growth.
2. Obtain a generation schedule, neglecting transmission losses, from the incremental cost slide rule.
3. With the generation schedule from step 2, determine penalty factors for all steam plants from the penalty-factor computer. This operation is accomplished by setting into the computer the various values of steam generation, as determined from step 2, as well as the hydro generation and foreign interchange flow.
4. Displace the incremental-fuel-rate strip of a given plant by the logarithm of the penalty factor. Determine a new generation schedule from the incremental-rate slide rule.
5. With the generation schedule determined from step 4, determine new penalty factors on the penalty-factor computer.
6. Using the new penalty factors from step 5 on the incremental slide rule, determine a new generation schedule.

The above rapidly converging cycle of calculation is repeated until the penalty factors used on the slide rule correspond to the generation schedule obtained from the slide rule. The rate of convergence can be increased by using a penalty factor for a given plant in steps 4 and 6 equal to the average of the old and new penalty factors for that plant. Thus in step 6 the penalty factors to be set on the slide rule for a given plant would be equal to the average of the penalty factors for that plant obtained in steps 3 and 5.

In the course of calculating the daily load schedule it is not necessary to start with step 2 each time. The usual procedure is to obtain an initial generation schedule from the incremental-rate slide rule, using the penalty factors corresponding to the generation schedule of the preceding hour.

This procedure converges quickly to the correct solution primarily for the following reasons:

1. The penalty factors are corrections on the incremental fuel costs, with the incremental fuel costs generally being the most important factor in determining allocation of generation.
2. Also, in determining generation schedules on an hour-to-hour basis only small changes in generation are involved, with correspondingly smaller changes in penalty factors.

The use of the penalty-factor computer in combination with the incremental slide rule by the American Gas and Electric Service Corporation has provided several specific advantages:

1. A means is provided of determining penalty factors rapidly for any of the various operating conditions which are experienced daily on a large integrated power system.
2. Savings are afforded in economic scheduling of plant generation by consideration of transmission losses.
3. A rapid means of evaluating incremental transmission losses incurred in transactions with interconnected companies is provided.
4. The computations previously required in precalculating penalty factors or generation schedules are eliminated.

This combination has proved to be a practical solution to the problem of system operating economy. The computer and the slide rule provide flexibility in minute-to-minute and hour-to-hour scheduling by permitting the exercise of judgment in tempering the absolute fuel economy to take into account such factors as distribution of spinning reserve, local area protection, voltage support, and temporary limitations of turbine and boiler equipment.

More recently, during the summer of 1957, an improved design version of the penalty-factor computer manufactured by the General Electric Company was placed in service on the Pacific Gas and Electric System. This computer for a twenty-four-source system is shown in Figure 6.15. The circuitry is indicated in Figure 6.16 for a two-source system.[6] The input powers are represented by a-c voltages. The voltage proportional to the given input power is adjusted by means of a tap switch calibrated in 100-mw steps and a variable transformer which interpolates between these steps. This voltage is fed through a reversing switch to an isolated transformer with a center tap secondary. The potentiometers for positive loss-formula coefficients are placed across one side of the secondary; the potentiometers for negative loss-formula coefficients, across the other side of the secondary. The desired $2B_{mn}P_n$ sum is fed into a servomechanism positioning unit. This unit consists of an a-c amplifier which drives a motor mechanically coupled to a potentiometer and pointer. It adjusts itself so that the feedback voltage is equal to the desired $2B_{mn}P_n$ sum. Position of the pointer then corresponds to the value of incremental loss. The scale calibration accomplishes the $1/(1 - 2B_{mn}P_n)$ operation. Since the excitation for the multitap transformer and for the feedback voltage from the indicator comes from the same source, it is not necessary to regulate the line voltage to this computer.

210   ECONOMIC OPERATION OF POWER SYSTEMS

Figure 6.15.   Pacific Gas and Electric Penalty-Factor Computer.

Figure 6.15 shows the twenty-four-source power panels located on the slanting front. The twenty-four push-button selectors for the various penalty factors are located on the writing surface. The computer is 45 in. high, 30 in. deep, and 60 in. long. This computer is designed so that by addition of appropriate input and output servomechanism amplifiers the computer will continuously accept telemetered source loadings and calculate penalty factors for all sources. The penalty-factor computer is then suitable for incorporation in the General Electric Automatic Dispatching System. This arrangement is described fully in *Economic Control of Interconnected Systems* by L. K. Kirchmayer.

OTHER LOAD-DISPATCHING COMPUTERS

A number of the manual steps involved in the penalty-factor computer and slide rule approach can be eliminated by the use of electrical analogues of the incremental production-cost characteristics. Consider,

CALCULATION OF GENERATION SYSTEM SCHEDULES 211

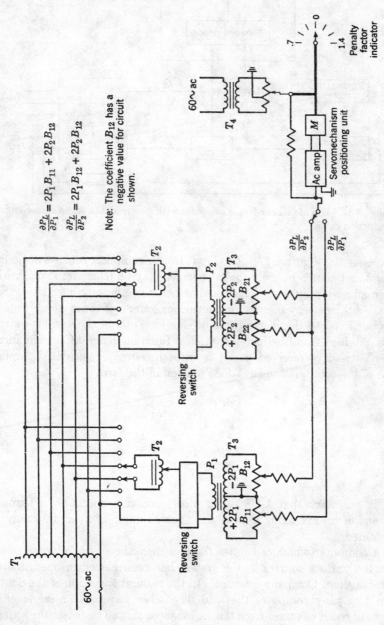

Figure 6.16. Schematic diagram of two-source penalty-factor computer.

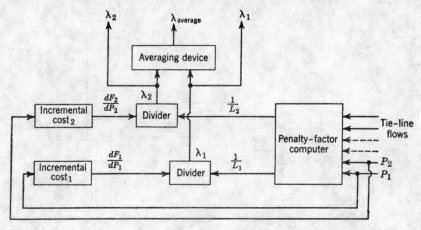

Figure 6.17. Use of penalty-factor computer with electrical analogues of incremental production costs.

for example, Figure 6.17. The input to this computer consists of tie-line flows and source outputs, both of which may be telemetered quantities or manually adjusted. The penalty-factor computer continuously calculates the penalty factors for the various sources of the system. Signals proportional to the incremental production costs of the various sources are obtained from the incremental-cost function generators. The incremental cost of received power from each source is calculated through use of the coordination equations written in the form

$$\frac{dF_1}{dP_1}\frac{1}{1/L_1} = \lambda_1$$

$$\frac{dF_2}{dP_2}\frac{1}{1/L_2} = \lambda_2$$

(.-29)

It is necessary that $\lambda_1 = \lambda_2$ for most economic operation. Through means of an averaging device, the average value of λ_1 and λ_2 may be obtained.

A computer which calculates the incremental cost of delivered power for the various sources in the system has been placed in operation on the Southern Company System.[7] In the manual operation of the Southern Company computer the load dispatcher inspects the incremental costs of received power from the various sources and instructs the sources with low values of λ to increase generation. Similarly, the sources with high values of λ are instructed to decrease generation. Desired load changes are teletyped to the dispatching offices of the four operating companies which comprise the Southern Company. These dispatching offices in turn issue orders to the plants for such load changes. When the load change has been completed it is confirmed to the central office containing the computer by teletype.

The General Electric Company has developed an economic dispatch computer which utilizes the principles illustrated in the system of Figure 6.5. An artist's sketch of this computer for a large integrated system is shown in Figure 6.18. The B_{mn} matrix is a field of miniature po-

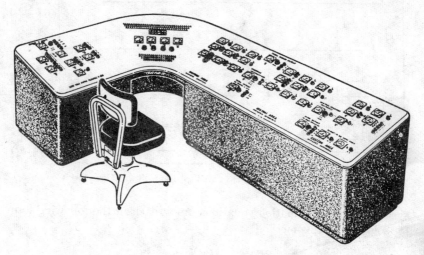

Figure 6.18. General Electric Economic Dispatching Computer.

tentiometers. The incremental-cost characteristic is represented by solid-state diode function generators. Each unit is treated separately and then combined with the other unit outputs within a given station to give the station power. A servomechanism is used to drive the value of λ such that the total generation obtained from the computer is equal to the desired total generation. The use of solid-state circuitry results in high reliability and requires a minimum of space and cooling equipment. The system is based upon plug-in-modules that allow for easy expansion and maintenance. The output is presented in digital form on a digital voltmeter and may also be presented optionally in printed form.

The Goodyear Aircraft Corporation, together with the Ohio Edison Company,[8] has adapted components from its line of electronic differential analyzers to obtain a dispatching computer which also functions in a manner similar to that shown in Figure 6.5.

The method of solution used for the West Penn Power Company computer [9,10] is indicated schematically in Figure 6.19 for a two-source

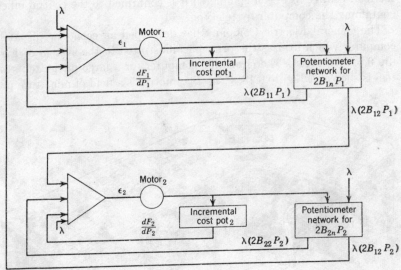

Figure 6.19. Schematic representation of West Penn Power Company Computer.

system. The coordination equations are written in the following form for solution by this method:

$$\frac{dF_1}{dP_1} + \lambda \frac{\partial P_L}{\partial P_1} - \lambda = \epsilon_1$$

$$\frac{dF_2}{dP_2} + \lambda \frac{\partial P_L}{\partial P_2} - \lambda = \epsilon_2$$

(6-30)

Expressing the incremental losses through means of a loss formula, we obtain

$$\frac{dF_1}{dP_1} + \lambda(2B_{11}P_1) + \lambda(2B_{12}P_2) - \lambda = \epsilon_1$$

$$\frac{dF_2}{dP_2} + \lambda(2B_{12}P_1) + \lambda(2B_{22}P_2) - \lambda = \epsilon_2$$

(6-31)

In Figure 6.19 the shaft position of motor 1 corresponds to P_1; the shaft position of motor 2, to P_2. Shaft position 1 is used to drive a potentiometer which generates dF_1/dP_1 in addition to a group of potentiom-

CALCULATION OF GENERATION SYSTEM SCHEDULES

eters which performs the operation $2\lambda B_{1n}P_n$. A similar process occurs for motor 2. Motors 1 and 2 are driven by errors ϵ_1 and ϵ_2, respectively, until the coordination equations are satisfied. The value of λ may be automatically adjusted by means of a comparison device in order to obtain a total computed generation equal to a desired value.

In the manual application of this computer the dispatcher compares the power indicated by each station shaft position with the power the station is actually carrying and issues orders to change those that are different.

ECONOMIC AUTOMATION [11, 12, 13, 14, 15, 16]

Important opportunities for additional savings are available through the use of an automatic dispatching system which provides economic automation of the power system. These savings result from

1. Improved fuel economy.
   (a) The dispatching system calculates the desired generator outputs for the conditions of the power system as they actually exist rather than for forecasted or preconceived conditions.
   (b) The dispatching system automatically executes the schedule, thereby resulting in closer adherence to proper allocation of generation.
2. Decrease in number of man-hours of time required to operate the system economically.

Various methods of economic automation are discussed in *Economic Control of Interconnected Systems* by L. K. Kirchmayer.

6.9 SUMMARY

The solution of the coordination equations may be practically effected by both analogue and digital methods.

The use of an iterative approach with an automatic digital computer is particularly valuable in precalculating schedules and in undertaking special studies. Since the computer program is general, a single routine is maintained in the computer-program library, which will permit scheduling of any size system in a practical manner. In addition to printing the allocation of generation, the digital computer also presents in printed form the incremental cost of received power, total transmission losses, received load, unit fuel input, and total system fuel input. The last four quantities are most difficult to calculate with generally available analogue devices.

Analogue computers, because of their lower initial cost, have been found more applicable as generation scheduling computers for installation in the dispatching office.

References

1. Coordination of Fuel Cost and Transmission Loss by Use of the Network Analyzer to Determine Plant Loading Schedules, E. E. George, H. W. Page, J. B. Ward. *AIEE Trans.*, Vol. 68, Part II, 1949, pp. 1152–1160.
2. *Tensor Analysis of Networks*, G. Kron. John Wiley and Sons, New York, 1939.
3. Automatic Digital Computer Applied to Generation Scheduling, A. F. Glimn, L. K. Kirchmayer, R. Habermann, Jr., R. W. Thomas. *AIEE Trans.*, Vol. 73, Part III, 1954, pp. 1267–1274.
4. A Transmission-Loss Penalty-Factor Computer, C. A. Imburgia, L. K. Kirchmayer, G. W. Stagg. *AIEE Trans.*, Vol. 73, Part III-A, 1954, pp. 567–570.
5. Design and Application of Penalty-Factor Computer, C. A. Imburgia, G. W. Stagg, L. K. Kirchmayer, K. R. Geiser. Proceedings of the American Power Conference, 1955, Vol. XVII, pp. 687–697.
6. Penalty-Factor Computers, H. H. Chamberlain, K. N. Burnett. Paper presented at the Annual Conference of the Engineering and Operation Section of PCEA, Los Angeles, California, March 21 and 22, 1957.
7. An Incremental Cost of Power-Delivered Computer, E. D. Early, W. E. Phillips, W. T. Shreve. *AIEE Trans.*, Vol. 74, Part III, 1955, pp. 529–535.
8. A Computer for Economic Scheduling and Control of Power Systems, C. D. Morrill, J. A. Blake. *AIEE Trans.*, Vol. 74, Part III, 1955, pp. 1136–1142.
9. Loss Evaluation—Part V: Economic Dispatch Computer—Design, R. B. Squires, R. T. Byerly, H. W. Colborn, W. R. Hamilton. *AIEE Trans.*, Vol. 75, Part III, 1956, pp. 719–727.
10. Loss Evaluation—Part IV: Economic Dispatch Computer Principles and Application, W. H. Osterle, E. L. Harder. *AIEE Trans.*, Vol. 75, Part III, 1956, pp. 387–394.
11. A New Automatic Dispatching System for Electric Power Systems, K. N. Burnett, D. W. Halfhill, B. R. Shepard. *AIEE Trans.*, Vol. 75, Part III, 1956, pp. 1049–1056.
12. A New Type Automatic Dispatching System at Kansas City, D. H. Cameron, E. L. Mueller. *ASME Technical Paper No. 56-A-215*.
13. "The Early Bird" Goes Automatic, E. J. Kompass. *Control Engineering*, December 1956, pp. 77–83.
14. Automatic Economic Dispatching and Load Control—Ohio Edison System, R. H. Travers. *AIEE Paper 57-143* presented at the Winter General Meeting, January 1957.
15. Economic Aspects of General Electric Automatic Dispatching System at Kansas City, D. H. Cameron, E. L. Mueller. *AIEE Conference Paper 57-144* presented at the Winter Meeting, January 1957.
16. Automatic Operation of Interconnected Areas, H. H. Chamberlain, A. F. Glimn, L. K. Kirchmayer. *AIEE Conference Paper* presented at the Summer General Meeting, June 1956.

Problems

Problem 6.1

Assume that the plant incremental-production-cost data may be approximated by

$$\frac{dF_1}{dP_1} = 2 + (1.0)P_1 = F_{11}P_1 + f_1$$

$$\frac{dF_2}{dP_2} = 1.5 + (1.0)P_2 = F_{22}P_2 + f_2$$

where P is expressed in per unit on a 100-mva base. Assume that both units are in operation and the maximum loading of each unit is 100 mw and that the minimum loading is 20 mw.

Assume that the B_{mn} coefficients in per unit on a 100-mva base are given by

	1	2
1	+0.10	−0.05
2	−0.05	+0.20

For $\lambda = 2.5$ solve the exact coordination equation

$$\frac{dF_n}{dP_n} + \lambda \frac{\partial P_L}{\partial P_n} = \lambda$$

using (1) the matrix inversion method and (2) the iterative method. Repeat for $\lambda = 2.0$ using the iterative method.

Problem 6.2

In order to determine whether to include transmission-loss considerations in the operation of a power system it is prudent to determine the amount of annual savings involved. As an example of the economic importance of transmission-loss considerations, determine the amount of these annual savings for the data of problems 2.1, 2.2, and 5. The load duration curve is given in Figure 6.20. Consider the case for which both units are in operation.

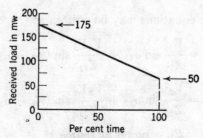

Figure 6.20. Load-duration curve for problem 6.2.

7 TRANSMISSION LOSSES AS A FUNCTION OF VOLTAGE PHASE ANGLE

7.1 INTRODUCTION

For simple circuit configurations, incremental losses and changes in total losses may be rigorously and simply expressed in terms of functions of voltage phase angles, driving point and transfer impedances, and voltage magnitudes. In certain limiting cases voltage magnitudes and driving-point and transfer impedances cancel out, leaving an expression involving X/R ratios and differences in voltage phase angles.

7.2 TWO-MACHINE SYSTEM WITHOUT INTERMEDIATE LOADS

The system under consideration is indicated in Figure 7.1. It is assumed that

1. Voltage magnitudes remain constant.
2. Reactive power flows in such a manner as to maintain constant voltage.

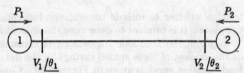

Figure 7.1. Two-machine system without intermediate loads.

The power-angle equations may be written as [1,2]

$$P_1 = \frac{V_1^2}{Z_{11}} \sin \alpha_{11} + \frac{V_1 V_2}{Z_{12}} \sin(\theta_{12} - \alpha_{12}) \qquad (7\text{--}1)$$

$$P_2 = \frac{V_2^2}{Z_{22}} \sin \alpha_{22} + \frac{V_2 V_1}{Z_{21}} \sin(\theta_{21} - \alpha_{21}) \qquad (7\text{--}2)$$

where
P_1 = power at source 1
P_2 = power at source 2

LOSSES AS FUNCTION OF VOLTAGE PHASE ANGLE

$Z_{11}, Z_{12}, Z_{21}, Z_{22}$ = absolute values of driving point and transfer impedances

$$\alpha_{11} = \tan^{-1} \frac{R_{11}}{X_{11}}$$

$$\alpha_{21} = \alpha_{12} = \tan^{-1} \frac{R_{12}}{X_{12}}$$

$$\alpha_{22} = \tan^{-1} \frac{R_{22}}{X_{22}}$$

V_1 = absolute value of voltage at source 1

V_2 = absolute value of voltage at source 2

$\theta_{12} = \theta_1 - \theta_2$

θ_1 = angle of voltage 1

θ_2 = angle of voltage 2

If the line-charging of the transmission line is lumped with the var requirements of the machine and if there are no intermediate loads or generators, then

$$Z_{11} = Z_{12} = Z_{21} = Z_{22}$$

$$\alpha_{11} = \alpha_{12} = \alpha_{21} = \alpha_{22}$$

It is intended to calculate the change in losses involved when the generation is swung between sources 1 and 2 by increasing the output of source 1 and decreasing the output of source 2.

CHANGE IN TOTAL LOSSES

The transmission losses are given by

$$\begin{aligned} P_L &= P_1 + P_2 \\ &= \frac{V_1^2}{Z_{11}} \sin \alpha_{11} + \frac{V_1 V_2}{Z_{12}} \sin(\theta_{12} - \alpha_{12}) \\ &\quad + \frac{V_2^2}{Z_{22}} \sin \alpha_{22} + \frac{V_2 V_1}{Z_{21}} \sin(\theta_{21} - \alpha_{21}) \end{aligned} \quad (7\text{-}3)$$

Assume that the system has changed to a new condition in which the angle between V_1 and V_2 increases to θ_{12}'. Then

$$P_L' = P_1' + P_2'$$

$$= \frac{V_1^2}{Z_{11}} \sin \alpha_{11} + \frac{V_1 V_2}{Z_{12}} \sin (\theta_{12}' - \alpha_{12})$$

$$+ \frac{V_2^2}{Z_{22}} \sin \alpha_{22} + \frac{V_2 V_1}{Z_{21}} \sin (\theta_{21}' - \alpha_{21}) \qquad (7\text{-}4)$$

The change in total losses is given by

$$\Delta P_L = P_L' - P_L = \frac{V_1 V_2}{Z_{12}} [\sin (\theta_{12}' - \alpha_{12}) - \sin (\theta_{12} - \alpha_{12})]$$

$$+ \frac{V_2 V_1}{Z_{21}} [\sin (\theta_{21}' - \alpha_{21}) - \sin (\theta_{21} - \alpha_{21})]$$

Recalling that

$$\sin (\alpha - \beta) = \sin \alpha \cos \beta - \sin \beta \cos \alpha$$

$$\Delta P_L = \frac{V_1 V_2}{Z_{12}}$$

$$\times [\sin \theta_{12}' \cos \alpha_{12} - \cos \theta_{12}' \sin \alpha_{12} - \sin \theta_{12} \cos \alpha_{12} + \cos \theta_{12} \sin \alpha_{12}]$$

$$+ \frac{V_2 V_1}{Z_{21}}$$

$$\times [-\sin \theta_{12}' \cos \alpha_{12} - \cos \theta_{12}' \sin \alpha_{12} + \sin \theta_{12} \cos \alpha_{12} + \cos \theta_{12} \sin \alpha_{12}]$$

$$= \frac{V_1 V_2}{Z_{12}} [2 (\cos \theta_{12} - \cos \theta_{12}') \sin \alpha_{12}] \qquad (7\text{-}5)$$

CALCULATION OF INCREMENTAL LOSS

The incremental loss which we are considering is given by dividing the change in loss by the change in generation of a given source when swinging generation between that source and one other source. For the system of Figure 7.1,

$$\frac{dP_{L1.2}}{dP_1} = \frac{\Delta P_{L1.2}}{\Delta P_1} = \begin{array}{l}\text{change in transmission loss divided by change} \\ \text{in generation at source 1 when swinging} \\ \text{generation between source 1 and source 2.}\end{array}$$

LOSSES AS FUNCTION OF VOLTAGE PHASE ANGLE

Also,

$$\frac{dP_{L1,2}}{dP_2} = \frac{\Delta P_{L1,2}}{\Delta P_2} = \begin{array}{l}\text{change in transmission loss divided by change} \\ \text{in generation at source 2 when swinging} \\ \text{generation between source 1 and source 2.}\end{array}$$

From equations 7–1 and 7–2 it is seen that for very small changes from a given operating condition,

$$\Delta P_1 = \frac{V_1 V_2}{Z_{12}} \cos(\theta_{12} - \alpha_{12}) \Delta \theta_{12} = \Psi_{12} \Delta \theta_{12} \qquad (7\text{–}6)$$

$$\Delta P_2 = \frac{V_2 V_1}{Z_{21}} \cos(\theta_{21} - \alpha_{21}) \Delta \theta_{21} = \Psi_{21} \Delta \theta_{21} \qquad (7\text{–}7)$$

where

$$\Psi_{12} = \frac{V_1 V_2}{Z_{12}} \cos(\theta_{12} - \alpha_{12}) \qquad (7\text{–}8)$$

$$\Psi_{21} = \frac{V_2 V_1}{Z_{21}} \cos(\theta_{21} - \alpha_{21}) \qquad (7\text{–}9)$$

The change in loss is then given by

$$\begin{aligned}\Delta P_{L1,2} &= \Delta P_1 + \Delta P_2 \\ &= \Psi_{12} \Delta \theta_{12} + \Psi_{21} \Delta \theta_{21} \\ &= (\Psi_{12} - \Psi_{21}) \Delta \theta_{12}\end{aligned} \qquad (7\text{–}10)$$

Then

$$\frac{dP_{L1,2}}{dP_1} = \frac{\Delta P_{L1,2}}{\Delta P_1} = \frac{(\Psi_{12} - \Psi_{21})}{\Psi_{12}} \qquad (7\text{–}11)$$

The above expression may be further simplified.

$$\begin{aligned}\frac{\Psi_{12} - \Psi_{21}}{\Psi_{12}} &= \frac{\cos(\theta_{12} - \alpha_{12}) - \cos(\theta_{21} - \alpha_{21})}{\cos(\theta_{12} - \alpha_{12})} \\ &= \frac{\cos\theta_{12}\cos\alpha_{12} + \sin\theta_{12}\sin\alpha_{12} - \cos\theta_{12}\cos\alpha_{12} + \sin\theta_{12}\sin\alpha_{12}}{\cos\theta_{12}\cos\alpha_{12} + \sin\theta_{12}\sin\alpha_{12}} \\ &= \frac{2\sin\theta_{12}\sin\alpha_{12}}{\cos\theta_{12}\cos\alpha_{12} + \sin\theta_{12}\sin\alpha_{12}}\end{aligned}$$

Dividing numerator and denominator by $\cos \theta_{12} \cos \alpha_{12}$, we obtain

$$\frac{\Psi_{12} - \Psi_{21}}{\Psi_{12}} = \frac{2 \tan \theta_{12} \tan \alpha_{12}}{1 + \tan \theta_{12} \tan \alpha_{12}}$$

Thus
$$\frac{dP_{L1.2}}{dP_1} = \frac{2 \tan \theta_{12}}{X_{12}/R_{12} + \tan \theta_{12}} \quad (7\text{--}12)$$

since
$$\tan \alpha_{12} = \frac{R_{12}}{X_{12}}$$

Similarly,
$$\frac{dP_{L1.2}}{dP_2} = \frac{-2 \tan \theta_{12}}{X_{12}/R_{12} - \tan \theta_{12}} \quad (7\text{--}13)$$

Typical results for $dP_{L1.2}/dP_1$ as a function of θ_{12} are plotted in Figure 7.2. The above equations, 7-12 and 7-13, correspond to those given

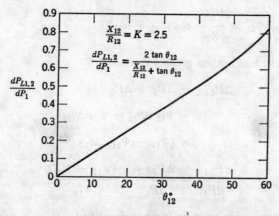

Figure 7.2. Plot of $dP_{L1.2}/dP_1$ as function of θ_{12} for system of Figure 7.1.

by Brownlee in Reference 3 if X_{12}/R_{12} is replaced by a quantity K equal to the X/R ratio of the line between source 1 and source 2.

Reference 3 suggests that the effect of an intermediate source may be approximated by the expression

$$\frac{dP_{L1.2}}{dP_1} = \frac{4K \tan \tfrac{1}{2}\theta_{12}}{(K + \tan \tfrac{1}{2}\theta_{12})^2} \quad (7\text{--}14)$$

if K is taken as the X/R ratio of the impedance between source 1 and source 2 with all other sources and loads open-circuited.

7.3 SYSTEM WITH INTERMEDIATE LOAD OR GENERATION

In this section we shall derive a rigorous expression for $dP_{L1,2}/dP_1$ for the case in which there is an intermediate load or source as indicated in Figure 7.3. It is assumed that the reactive characteristics of this intermediate source or load are such as to maintain constant voltage. We shall designate by the subscript 3 the quantities relating to this intermediate point. The network connecting these three points is not restricted in any manner as to its configuration.

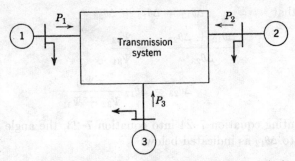

Figure 7.3. Schematic representation of system with intermediate load or generation.

For this case, we have

$$P_1 = \frac{V_1^2}{Z_{11}} \sin \alpha_{11} + \frac{V_1 V_2}{Z_{12}} \sin (\theta_{12} - \alpha_{12}) + \frac{V_1 V_3}{Z_{13}} \sin (\theta_{13} - \alpha_{13}) \tag{7-15}$$

$$P_2 = \frac{V_2^2}{Z_{22}} \sin \alpha_{22} + \frac{V_2 V_1}{Z_{21}} \sin (\theta_{21} - \alpha_{21}) + \frac{V_2 V_3}{Z_{23}} \sin (\theta_{23} - \alpha_{23}) \tag{7-16}$$

$$P_3 = \frac{V_3^2}{Z_{33}} \sin \alpha_{33} + \frac{V_3 V_1}{Z_{31}} \sin (\theta_{31} - \alpha_{31}) + \frac{V_3 V_2}{Z_{32}} \sin (\theta_{32} - \alpha_{32}) \tag{7-17}$$

Then

$$\Delta P_1 = \Psi_{12} \Delta \theta_{12} + \Psi_{13} \Delta \theta_{13} \tag{7-18}$$

$$\Delta P_2 = \Psi_{21} \Delta \theta_{21} + \Psi_{23} \Delta \theta_{23} \tag{7-19}$$

$$\Delta P_3 = \Psi_{31} \Delta \theta_{31} + \Psi_{32} \Delta \theta_{32} \tag{7-20}$$

where

$$\Psi_{jk} = \frac{V_j V_k}{Z_{jk}} \cos (\theta_{jk} - \alpha_{jk}) \text{ with } j \neq k \tag{7-21}$$

In swinging generation between sources 1 and 2 the power P_3 is to remain

224   ECONOMIC OPERATION OF POWER SYSTEMS

constant. Thus,
$$\Delta P_3 = 0 \tag{7-22}$$

This relation is used to express $\Delta\theta_{13}$ and $\Delta\theta_{23}$ in terms of $\Delta\theta_{12}$. From equation 7–20 we obtain

$$\Psi_{31}\,\Delta\theta_{31} = -\Psi_{32}\,\Delta\theta_{32}$$

$$\frac{\Delta\theta_{31}}{\Delta\theta_{32}} = \frac{\Delta\theta_{13}}{\Delta\theta_{23}} = -\frac{\Psi_{32}}{\Psi_{31}}$$

Recalling that
$$\Delta\theta_{13} = \Delta\theta_{12} + \Delta\theta_{23} \tag{7-23}$$

we obtain
$$\frac{\Delta\theta_{12} + \Delta\theta_{23}}{\Delta\theta_{23}} = -\frac{\Psi_{32}}{\Psi_{31}}$$

$$\Delta\theta_{23} = \Delta\theta_{12}\,\frac{-\Psi_{31}}{\Psi_{32} + \Psi_{31}} \tag{7-24}$$

By substituting equation 7–24 into equation 7–23, the angle $\Delta\theta_{13}$ may be related to $\Delta\theta_{12}$ as indicated below:

$$\Delta\theta_{13} = \Delta\theta_{12}\left(\frac{\Psi_{32}}{\Psi_{32} + \Psi_{31}}\right) \tag{7-25}$$

As before

$$\Delta P_{L1,2} = \Delta P_1 + \Delta P_2$$
$$= \Psi_{12}\,\Delta\theta_{12} + \Psi_{13}\,\Delta\theta_{13} + \Psi_{21}\,\Delta\theta_{21} + \Psi_{23}\,\Delta\theta_{23}$$
$$= (\Psi_{12} - \Psi_{21})\,\Delta\theta_{12} + \Psi_{13}\,\Delta\theta_{13} + \Psi_{23}\,\Delta\theta_{23} \tag{7-26}$$

Substituting equation 7–24 and 7–25 into 7–26,

$$\Delta P_{L1,2} = \left[(\Psi_{12} - \Psi_{21}) + \left(\frac{\Psi_{13}\Psi_{32}}{\Psi_{32} + \Psi_{31}}\right) - \left(\frac{\Psi_{31}\Psi_{23}}{\Psi_{32} + \Psi_{31}}\right)\right]\Delta\theta_{12} \tag{7-27}$$

Substituting equation 7–25 into 7–18

$$\Delta P_1 = \left[\Psi_{12} + \frac{\Psi_{13}\Psi_{32}}{\Psi_{32} + \Psi_{31}}\right]\Delta\theta_{12} \tag{7-28}$$

Then

$$\frac{dP_{L1,2}}{dP_1} = \frac{\Delta P_{L1,2}}{\Delta P_1} = \frac{(\Psi_{12} - \Psi_{21}) + \dfrac{\Psi_{13}\Psi_{32}}{\Psi_{32} + \Psi_{31}} - \dfrac{\Psi_{31}\Psi_{23}}{\Psi_{32} + \Psi_{31}}}{\Psi_{12} + \dfrac{\Psi_{13}\Psi_{32}}{\Psi_{32} + \Psi_{31}}} \tag{7-29}$$

LOSSES AS FUNCTION OF VOLTAGE PHASE ANGLE

The procedure involved for a system of n voltage-supported points is indicated in Appendix 7.

The foregoing expression may be simplified if the three-point system assumes the particular form shown in Figure 7.4. For this particular configuration

$$Z_{12} = Z_{21} = \infty$$

$$Z_{13} = Z_{31} = \text{impedance of line 1-3}$$

$$Z_{23} = Z_{32} = \text{impedance of line 2-3}$$

Since $Z_{12} = Z_{21} = \infty$, $\Psi_{12} = \Psi_{21} = 0$

Equation 7-29 then reduces to

$$\frac{dP_{L1.2}}{dP_1} = 1 - \frac{\Psi_{23} \Psi_{31}}{\Psi_{32} \Psi_{13}} \tag{7-30}$$

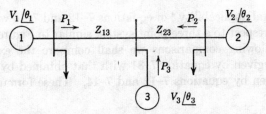

Figure 7.4. Simplified three-point system.

From equation 7-21 and 7-30

$$\frac{\Psi_{23} \Psi_{31}}{\Psi_{32} \Psi_{13}} = \frac{\cos(\theta_{23} - \alpha_{23}) \cos(\theta_{31} - \alpha_{31})}{\cos(\theta_{32} - \alpha_{32}) \cos(\theta_{13} - \alpha_{13})}$$

$$= \left[\frac{1 + \tan \theta_{23} \tan \alpha_{23}}{1 - \tan \theta_{23} \tan \alpha_{23}}\right]\left[\frac{1 - \tan \theta_{13} \tan \alpha_{13}}{1 + \tan \theta_{13} \tan \alpha_{13}}\right]$$

Then
$$\frac{dP_{L1.2}}{dP_1} = 1 - \frac{\left[\dfrac{X_{23}}{R_{23}} + \tan \theta_{23}\right]\left[\dfrac{X_{13}}{R_{13}} - \tan \theta_{13}\right]}{\left[\dfrac{X_{23}}{R_{23}} - \tan \theta_{23}\right]\left[\dfrac{X_{13}}{R_{13}} + \tan \theta_{13}\right]} \tag{7-31}$$

Assume that
$$\frac{X_{23}}{R_{23}} = \frac{X_{13}}{R_{13}} = K$$

$$\theta_{13} = \theta_{32} = \frac{\theta_{12}}{2}$$

Equation 7-31 then becomes

$$\frac{dP_{L1,2}}{dP_1} = 1 - \frac{\left(K - \tan\frac{\theta_{12}}{2}\right)\left(K - \tan\frac{\theta_{12}}{2}\right)}{\left(K + \tan\frac{\theta_{12}}{2}\right)\left(K + \tan\frac{\theta_{12}}{2}\right)} = \frac{4K\tan\frac{\theta_{12}}{2}}{\left(K + \tan\frac{\theta_{12}}{2}\right)^2}$$
(7-14)

Reference 3 suggests that the expression 7-14 may be used without consideration of

    a—Transfer impedance between plants
    b—System load distribution, or
    c—Generation of other plants

and that equation 7-12 may be used directly when the angle θ_{12} is less than 15°.

The general applicability [4] of equations 7-12 and 7-14 will be investigated with respect to a three-point system similar to Figure 7.4. In this and the following comparisons we shall compare the exact value of $dP_{L1,2}/dP_1$ given by equation 7-31 with that obtained by the approximations given by equations 7-12 and 7-14. These formulas are tabulated below:

$$\frac{dP_{L1,2}}{dP_1} = \frac{2\tan\theta_{12}}{K + \tan\theta_{12}} \tag{7-12}$$

$$\frac{dP_{L1,2}}{dP_1} = \frac{4K\tan\tfrac{1}{2}\theta_{12}}{(K + \tan\tfrac{1}{2}\theta_{12})^2} \tag{7-14}$$

$$\frac{dP_{L1,2}}{dP_1} = 1 - \frac{\left[\dfrac{X_{23}}{R_{23}} + \tan\theta_{23}\right]\left[\dfrac{X_{13}}{R_{13}} - \tan\theta_{13}\right]}{\left[\dfrac{X_{23}}{R_{23}} - \tan\theta_{23}\right]\left[\dfrac{X_{13}}{R_{13}} + \tan\theta_{13}\right]} \tag{7-31}$$

The effect of an intermediate load is illustrated by considering the system shown in Figures 7.5 and 7.6.

The quantity K_{13} is defined to be the X/R ratio of line 1–3; similarly, K_{23} is the X/R ratio of line 2–3. The ratio of K_{23} to K_{13} is given by n. The value of K for the transmission line impedance between 1 and 2 was maintained at a constant value of 2.5. Thus equations 7-12 and 7-14 are used with $K = 2.5$. The various curves illustrated correspond to various values of n; thus $n = 1$ corresponds to both lines having an X/R ratio of 2.5. Similarly, $n = 2$ corresponds to line 1–3 having an X/R ratio of 1.875 and line 2–3 having an X/R ratio of 3.75.

LOSSES AS FUNCTION OF VOLTAGE PHASE ANGLE

In Figure 7.5 the angle θ_{12} was maintained constant at 30°. It will be noted that θ_{12} is equal to $\theta_{13} - \theta_{23}$. With θ_{12} maintained constant at 30°, θ_{23} and θ_{13} were varied as indicated by the abscissa. We note that

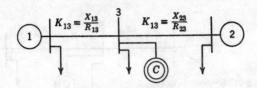

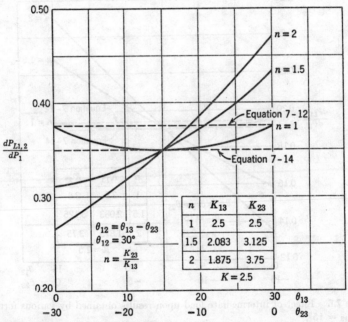

Figure 7.5. Effect of intermediate load upon results obtained by various formulas with $\theta_{12} = 30°$.

if the X/R ratios of both lines are the same the effect of the intermediate load and its relative angular position is not very great. However, if the X/R ratios of the lines are different, considerable discrepancies may occur. For example, with $n = 2$, $dP_{L1,2}/dP_1$ may vary from 0.27 to 0.47, instead of being constant at 0.35, as indicated by equation 7–14, or 0.375, as indicated by equation 7–12.

Figure 7.6 illustrates that similar differences occur when the angular difference θ_{12} is maintained constant at 15°; thus $dP_{L1,2}/dP_1$ may vary from 0.13 to 0.25 for $n = 2$ where the approximate answer given by equations 7–12 and 7–14 is approximately 0.19. Here we note that the

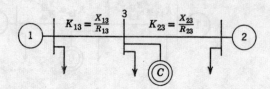

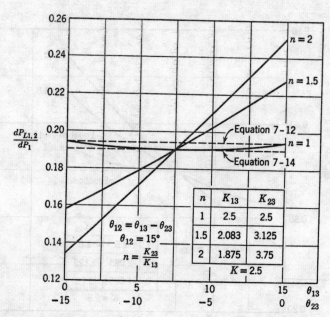

Figure 7.6. Effect of intermediate load upon results obtained by various formulas with $\theta_{12} = 15°$.

correct result is dependent upon the relative angular position of the intermediate load and the X/R ratio of the transfer impedances Z_{13} and Z_{23}, even though the K of the transmission line between 1 and 2 remains constant at 2.5.

The effect of intermediate plants is illustrated by the simple system shown in Figure 7.7.

In this case the angular difference θ_{12} is maintained at zero degrees and θ_{13} and θ_{23} are chosen equal to one another. Equations 7–12 and

7–14 applied to this case would indicate $dP_{L1,2}/dP_1$ to be identically zero for this condition. Inspection of Figure 7.7 indicates that if the X/R ratios of the two lines are identical $dP_{L1,2}/dP_1$ is zero. However,

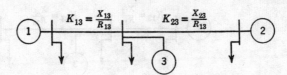

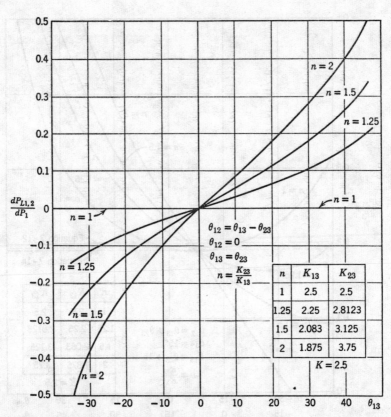

Figure 7.7. Effect of intermediate generation upon results obtained by various formulas with $\theta_{13} = \theta_{23}$.

if the X/R ratios of the lines are different, we note that the value of $dP_{L1,2}/dP_1$ may differ considerably from zero. For example, with $n = 2$ and $\theta_{13} = \theta_{23} = \pm 30°$, the value of $dP_{L1,2}/dP_1$ would be either $+0.28$ or -0.38, as compared to zero given by equations 7–12 and 7–14. These differences become much more pronounced as the X/R ratios of the individual lines become increasingly different. Such differences in X/R ratio occur because of different conductor sizes in the system and also because of transformers in the system.

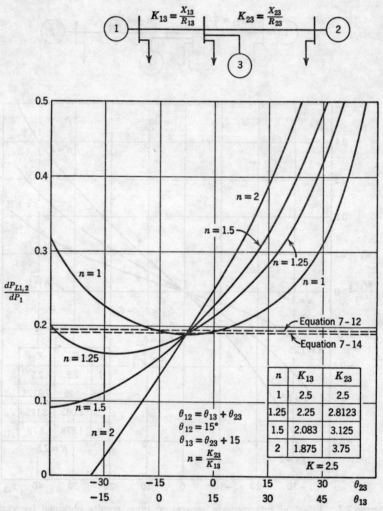

Figure 7.8. Effect of intermediate generation upon results obtained by various formulas with $\theta_{13} = \theta_{23} + 15°$.

Figure 7.8 also studies the effect of intermediate generation. In this case the angle θ_{13} is assumed to be 15° greater than θ_{23}. The results obtained by application of equations 7-12 and 7-14 are as indicated. The correct value of $dP_{L1,2}/dP_1$ is plotted as a function of θ_{23} and θ_{13} as indicated in Figure 7.8. It will be noted that under these conditions, even with X/R ratios of the two lines identical, there is a discrepancy between the results obtained by equation 7–12 and 7–14 and the correct result when the angular differences are large.

From the results presented in Figures 7.5 to 7.8 it appears that the use of the simplified expressions 7–12 and 7–14 yields a good approximation to incremental losses when the X/R ratios of the elements are very similar.

7.4 APPLICATION OF SIMPLIFIED PHASE-ANGLE FORMULAS

The paper entitled "The Determination of Incremental and Total Loss Formulas from Functions of Voltage Phase Angles," [5] by C. R. Cahn, presented a comparison of results obtained by application of a phase-angle formula with those obtained by a loss formula calculated in a manner similar to that described in Chapter 4. The latter formula is referred to as a B_{mn} formula. This comparison was made on a representation of the 110-kv transmission system of the Niagara Mohawk Power Corporation as indicated in Figure 7.9. The New York State Electric and Gas System has been reduced to a reactance equivalent, so that the losses in the NYSEG system are not present, but an approximation of the effect of the parallel lines of the NYSEG system upon losses in the Niagara Mohawk system is obtained.

The phase-angle formulas developed by Cahn parallel the derivations given in Reference 3 and Section 7.2, except that a new variable equal to the average of the sending and receiving end powers is introduced. Define this variable as

$$P_{1-2} = \frac{P_1 - P_2}{2} \qquad (7\text{–}32)$$

From equations 7–1 and 7–2 and for the system of Figure 7.1,

$$P_{1-2} = \frac{1}{2} \frac{V_1 V_2}{Z_{12}} [\sin(\theta_{12} - \alpha_{12}) - \sin(\theta_{21} - \alpha_{21})] \qquad (7\text{–}33)$$

Then
$$\Delta P_{1-2} = P_{1-2}' - P_{1-2}$$
$$= \frac{1}{2} \frac{V_1 V_2}{Z_{12}} [\sin(\theta_{12}' - \alpha_{12}) - \sin(\theta_{21}' - \alpha_{21})$$
$$- \sin(\theta_{12} - \alpha_{12}) + \sin(\theta_{21} - \alpha_{21})]$$

Recalling that $\sin(\alpha - \beta) = \sin\alpha\cos\beta + \sin\beta\cos\alpha$,

$$\Delta P_{1-2} = \frac{V_1 V_2}{Z_{12}} \cos\alpha_{12} [(\sin\theta_{12}' - \sin\theta_{12})] \qquad (7\text{–}34)$$

232 ECONOMIC OPERATION OF POWER SYSTEMS

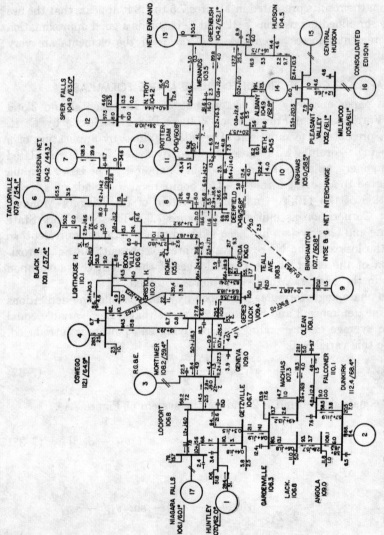

Figure 7.9. Impedance diagram and base case load flows for Niagara Mohawk 110-kv system. Flows on mw and megavars. Impedances in per cent on 100-mva base.

LOSSES AS FUNCTION OF VOLTAGE PHASE ANGLE

Using the relation

$$\sin \alpha - \sin \beta = 2 \cos \tfrac{1}{2}(\alpha + \beta) \sin \tfrac{1}{2}(\alpha - \beta)$$

equation 7-34 becomes

$$\Delta P_{1-2} = \frac{2V_1 V_2}{Z_{12}} \cos \alpha_{12} \left[\cos \left(\frac{\theta_{12}' + \theta_{12}}{2} \right) \right] \left[\sin \left(\frac{\theta_{12}' - \theta_{12}}{2} \right) \right] \quad (7\text{-}35)$$

Using the relation

$$\cos \alpha - \cos \beta = -2 \sin \tfrac{1}{2}(\alpha + \beta) \sin \tfrac{1}{2}(\alpha - \beta)$$

equation 7-5 becomes

$$\Delta P_L = \frac{4V_1 V_2}{Z_{12}} \sin \alpha_{12} \left[\sin \frac{(\theta_{12}' + \theta_{12})}{2} \right] \left[\sin \frac{(\theta_{12}' - \theta_{12})}{2} \right] \quad (7\text{-}36)$$

Dividing equation 7-36 by 7-35, we obtain

$$\frac{\Delta P_L}{\Delta P_{1-2}} = 2 \tan \alpha_{12} \tan \left(\frac{\theta_{12}' + \theta_{12}}{2} \right) \quad (7\text{-}37)$$

or

$$\Delta P_L = \left[2 \frac{2}{K} \tan \left(\frac{\theta_{12}' + \theta_{12}}{2} \right) \right] \Delta P_{1-2} \quad (7\text{-}38)$$

As ΔP_{1-2} approaches zero, equation 7-37 becomes

$$\frac{dP_L}{dP_{1-2}} = 2 \frac{2}{K} \tan \theta_{12} \quad (7\text{-}39)$$

Compare this formula with equations 7-12 and 7-13.

Equation 7-38 is the formula which is used in the comparisons that follow.

The approximations involved in the determination of the B_{mn} loss formula used in the comparisons includes

1. Each load current remains a constant complex fraction of the total load current.
2. Generator voltage magnitudes remain at constant values.
3. Generator voltage angles remain fixed.
4. Generator Q/P rates remain fixed at their base case values.

In this study the accuracy of the formulas was evaluated by comparison of calculated formula values to actual $I^2 R$ values for cases obtained by swinging Oswego against other plants and also against various interconnections. The results obtained indicate that the best correlation between actual and calculated losses is given by the B_{mn} loss formula.

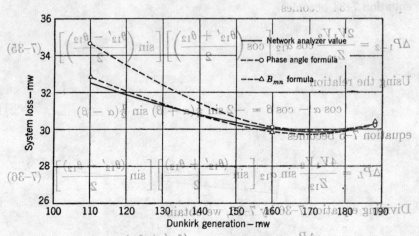

Figure 7.10. Comparison of actual and calculated losses for exchange of generation between Dunkirk and Oswego.

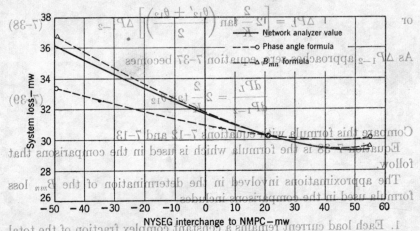

Figure 7.11. Comparison of actual and calculated losses for exchange of generation between NYSEG and Oswego.

The results obtained by phase-angle formula 7-38 appear to be in error, principally in the swings between Oswego and Dunkirk (Figure 7.10) and Oswego and the New York State Electric and Gas System (Figure 7.11). In the Dunkirk swing of approximately 80 mw the error in the change in loss given by the B_{mn} formula is negligible. The error by phase-angle formula 7–38 is 2.2 mw. For the 100-mw swing between Oswego and NYSEG it is seen that the error in the change in loss was 3.5 mw for phase-angle formula 7–38 and 0.4 mw for the B_{mn} formula. The errors in this phase-angle method result chiefly from the effects of different X/R ratios in the system and the effect of intermediate loads and generation when these different ratios exist. The B_{mn} formula does not rely on any assumptions with regard to the X/R ratios of the lines in the system. The errors illustrated by Mr. Cahn's study in the application of simplified phase-angle formulas, such as given by equations 7–5, 7–12, 7–13, 7–14, 7–36, 7–38, and 7–39, are usually significant in systems in which the transmission lines are at different voltage levels and in which transformers appear.

The results obtained by the use of phase-angle methods may be improved by application of formulas similar to equation 7–31; or, if equations of the form of 7–12 and 7–39 are used, it is necessary that the system be divided into parts, with the elements in each part having approximately the same X/R ratio.

7.5 SUMMARY

For transmission systems made up of elements with similar X/R ratios, the incremental loss obtained in swinging an increment of generation between points j and k may be closely approximated by the following simple expression:

$$\frac{dP_{Lj,k}}{dP_j} = \frac{2 \tan \theta_{jk}}{K + \tan \theta_{jk}} \tag{7-40}$$

 = change in transmission loss divided by change in generation at source j when swinging generation between source j and source k

θ_{jk} = angular difference between voltages of points j and k
$\phantom{\theta_{jk}}$ = $\theta_j - \theta_k$

K = X/R ratio of impedance between point j and point k with all other sources and loads open-circuited

APPENDIX 7. CONSIDERATION OF THE GENERAL CASE

The procedure involved in deriving an expression for $dP_{L1,2}/dP_1$ in the general case in which there are n intermediate sources and loads is outlined here. This analysis is restricted to the case in which all voltages are assumed to remain fixed in their absolute values. Here the expression for ΔP_1 and ΔP_2 is given by

$$\Delta P_1 = \sum_{k \neq 1}^{n} \frac{E_1 E_k}{Z_{1k}} \cos(\theta_{1k} - \alpha_{1k}) \Delta \theta_{1k} \tag{7-41}$$

$$\Delta P_2 = \sum_{k \neq 2}^{n} \frac{E_2 E_k}{Z_{2k}} \cos(\theta_{2k} - \alpha_{2k}) \Delta \theta_{2k} \tag{7-42}$$

For the remaining $n - 2$ points there are $n - 2$ equations of the form

$$\Delta P_j = \sum_{k \neq j}^{n} \frac{E_j E_k}{Z_{jk}} \cos(\theta_{jk} - \alpha_{jk}) \Delta \theta_{jk}, \quad j \neq 1, 2 \tag{7-43}$$

Assuming all ΔP_n to be zero but 1 and 2, we obtain the following $n - 2$ equations:

$$0 = \sum_{k \neq j}^{n} \frac{E_j E_k}{Z_{jk}} \cos(\theta_{jk} - \alpha_{jk}) \Delta \theta_{jk}, \quad i \neq 1, 2 \tag{7-44}$$

From the preceding equations and the relations

$$\Delta \theta_{1k} = \Delta \theta_{12} + \Delta \theta_{2k} \tag{7-45}$$

$$\Delta \theta_{mn} = \Delta \theta_{mp} + \Delta \theta_{pn} \tag{7-46}$$

the various angles $\Delta \theta_{jk}$ with $j, k \neq 1, 2$ may be expressed in terms of $\Delta \theta_{12}$.

As before, $\Delta P_{L1,2} = \Delta P_1 + \Delta P_2$. Dividing $\Delta P_{L1,2}$, by ΔP_1, an expression for $dP_{L1,2}/dP_1$ is obtained. The reader will note that this expression may be cumbersome to use.

References

1. *Power System Stability*, Vol. I, *Steady State Stability*, Seldon B. Crary. John Wiley and Sons, New York, 1945.
2. *Circuit Analysis of A-C Power Systems*, Vol. I, *Symmetrical and Related Components*, Edith Clarke. John Wiley and Sons, New York, 1943.
3. Coordination of Incremental Fuel Costs and Incremental Transmission Losses by Functions of Voltage Phase Angles, W. R. Brownlee. *AIEE Trans.*, Vol. 73, Part III, 1954, pp. 529–533.
4. Discussion of Reference 3 by L. K. Kirchmayer and A. F. Glimn. *AIEE Trans.*, Vol. 73, Part III, 1954, pp. 535–537.
5. The Determination of Incremental and Total Loss Formulas from Functions of Voltage Phase Angles, C. R. Cahn. *AIEE Trans.*, Vol. 74, Part III, 1955, pp. 161–176.

Problems

Problem 7

Calculate $dP_{L1,2}/dP_1$ by the approximate formulas 7–12 and 7–14 and the exact formula 7–31 for the system given in Figure 7.4 and the following data:

$$V_1/\theta_1 = 1.0/\underline{10°} \quad V_2/\theta_2 = 1.0/\underline{5°} \quad V_3/\theta_3 = 1.0/\underline{-5°}$$

$$Z_{13} = 0.05 + j.5 \quad Z_{23} = 0.25 + j.5$$

8 DESCRIPTION OF ALTERNATIVE COORDINATION METHODS

8.1 INTRODUCTION

Chapter 5, Coordination of Incremental Production Costs and Incremental Transmission Losses for Optimum Economy, discusses various methods of coordinating plant costs and transmission losses by the use of a total transmission-loss formula. This chapter discusses various alternative methods available for coordinating incremental plant production costs and incremental transmission losses.

8.2 SCHEDULING BY SUCCESSIVE TESTS ON NETWORK ANALYZER TO EVALUATE CHANGES IN SYSTEM INPUT

An interesting approach to the coordination problem involving the use of a network analyzer to obtain transmission losses in order to evaluate changes in system input has been suggested by T. W. Schroeder.[1] This procedure involves the following steps:

1. Set up the transmission system on the network analyzer in the same manner as for a load-flow study.
2. Choose a total system load such that all generating plants are at their minimum loads.
3. By means of the following tests the plant which can carry the next increment of load most economically is determined. First, increase the received load by an amount ΔP_R. The various loads on the analyzer are recalibrated to give this increase in total system load of ΔP_R.

Let plant 1 increase its output by an amount ΔP_1 to supply this increase in system load, the outputs of all other plants remaining constant. Read all scalar-line currents from the network analyzer.

Repeat the above test for all plants in turn.

4. Calculate the losses, line-by-line, for each of these tests. If there are n plants, n such tests are involved. Let ΔP_{L1} equal increase in losses when only plant 1 supplies the increase in system load. In general, let ΔP_{Ln} equal increase in losses when only plant n supplies the increase in

system load. This step is greatly expedited if $\sum_k i_k{}^2 R_k$ may be read directly off the analyzer.

5. Calculate the change in system input for each of the tests. The plant with the smallest change in dollar-per-hour input is obviously the plant to which this change in load should be allocated.

For example, the change in input to plant 1 is given by

$$\Delta F_1 = \frac{dF_1}{dP_1} \Delta P_1 = \frac{dF_1}{dP_1} (\Delta P_{L1} + \Delta P_R) = \Delta \text{ dollars per hour}$$

where dF_1/dP_1 = incremental production cost of plant 1. It is felt that $\Delta P_{L1} + \Delta P_R$ may be more accurately determined than direct measurement of ΔP_1. In a similar manner calculate ΔF_n for all plants.

6. From step 5 a new generation schedule is obtained. Steps 3, 4, and 5 are repeated successively until all plants are at their maximum loadings.

Certain practical considerations in the application of this method may make it difficult and time consuming to use:

1. Step 3 above would be difficult to execute because of the time involved in balancing generation on the analyzer so precisely as to take up the next block of generation only on one machine at a time. The use of automatic wattmetric control for the analyzer generators alleviates this problem.

2. Each point of the generation schedule requires recalibration of all loads.

3. Each point of the generation schedule requires as many $\sum_k I_k{}^2 R_k$ as there are variable plants.

4. The method is not flexible, since it would be necessary to reset the system on the analyzer and restudy if fuel costs or assumed number of units change. The methods discussed in Chapter 5 would only require the solution of a new set of equations.

This proposed method should be of value in conducting preliminary investigations into the effect of transmission losses upon the economic allocation of generation.

The calculation of the load flows through digital methods would eliminate a number of the difficulties outlined.

8.3 MODIFIED COORDINATION EQUATIONS INVOLVING DIRECT COMPARISON OF COSTS BETWEEN PLANTS

From Chapter 5 it will be recalled that the optimum generation schedule is obtained when the following set of coordination equations is satisfied:

$$\frac{dF_j}{dP_j}\frac{1}{\left(1-\frac{\partial P_L}{\partial P_j}\right)} = \lambda \qquad (8\text{-}1)$$

where $\dfrac{dF_j}{dP_j}$ = incremental production cost of plant j

$\dfrac{\partial P_L}{\partial P_j}$ = incremental transmission loss of plant j

λ = incremental cost of received power

In terms of words, equation 8–1 results in the incremental cost of received power being the same from each source. The comparison of the incremental costs for equation 8–1 is made at the hypothetical system load point.

If transmission loss is considered a function of all input power to the system,

$$dP_L = \sum_j \frac{\partial P_L}{\partial P_j} dP_j \qquad (8\text{-}2)$$

where P_j = powers entering the system (loads, interconnections, or plant generations).

When all dP_j are zero except for the two particular plants P_j and P_n,

$$dP_{Lj,n} = \frac{\partial P_L}{\partial P_j} dP_j + \frac{\partial P_L}{\partial P_n} dP_n \qquad (8\text{-}3)$$

The incremental loss $dP_{Lj,n}/dP_j$ is the limiting ratio of the change in loss to the change in generation at plant j when power is transferred from plant j to plant n, with all loads and other plant generations remaining fixed.

It is evident that

$$dP_{Lj,n} = dP_j + dP_n \qquad (8\text{-}4)$$

From equations 8–3 and 8–4, we obtain

$$dP_j\left(1 - \frac{\partial P_L}{\partial P_j}\right) + dP_n\left(1 - \frac{\partial P_L}{\partial P_n}\right) = 0 \qquad (8\text{-}5)$$

or

$$\frac{dP_j}{dP_n} = -\frac{\left(1 - \frac{\partial P_L}{\partial P_n}\right)}{\left(1 - \frac{\partial P_L}{\partial P_j}\right)} \qquad (8\text{-}6)$$

From equation 8-1 we may write

$$\frac{dF_j}{dP_j}\frac{1}{\left(1-\dfrac{\partial P_L}{\partial P_j}\right)} = \lambda = \frac{dF_n}{dP_n}\frac{1}{\left(1-\dfrac{\partial P_L}{\partial P_n}\right)} \qquad (8\text{-}7)$$

From equations 8-7, 8-6, and 8-4,

$$\frac{\dfrac{dF_n}{dP_n}}{\dfrac{dF_j}{dP_j}} = \frac{\left(1-\dfrac{\partial P_L}{\partial P_n}\right)}{\left(1-\dfrac{\partial P_L}{\partial P_j}\right)} = -\frac{dP_j}{dP_n} = -\frac{dP_j}{dP_{Lj,n}-dP_j}$$

$$\frac{\dfrac{dF_n}{dP_n}}{\dfrac{dF_j}{dP_j}} = \frac{1}{\left(1-\dfrac{dP_{Lj,n}}{dP_j}\right)} \qquad (8\text{-}8)$$

or
$$\frac{dF_n}{dP_n} = \frac{1}{\left(1-\dfrac{dP_{Lj,n}}{dP_j}\right)}\frac{dF_j}{dP_j} \qquad (8\text{-}9)$$

For a comparison between plants 1 and 2 with $j=1$ and $n=2$, we obtain:

$$\frac{dF_2}{dP_2} = \frac{1}{\left(1-\dfrac{dP_{L1,2}}{dP_1}\right)}\frac{dF_1}{dP_1} \qquad (8\text{-}10)$$

From this expression it is seen that the incremental cost of power at plant bus 2 is equal to the incremental cost of power at plant bus 1 corrected for the effect of the incremental transmission loss involved in swinging generation between plant 1 and plant 2.

Equation 8-10 may be determined [2] through physical reasoning by conducting a test on the system in which generation is arbitrarily swung from plant 1 to plant 2. If the load and all other plants are assumed to remain constant, then

$$\Delta P_1 + \Delta P_2 - \Delta P_{L1,2} = 0 \qquad (8\text{-}11)$$

Assume ΔP_1 is a positive number, i.e., the output of source 1 is increased over its initial amount. Then ΔP_2 will, of course, be a negative number. The change in transmission loss ($\Delta P_{L1,2}$) incurred because of this transfer of power between plants 1 and 2 will be either positive or negative or zero, depending upon the initial loading of the transmission system.

The input in dollars per hour to source 1 will then be increased by an amount ΔF_1; similarly, the input in dollars per hour to source 2 will then be decreased by an amount ΔF_2. The change in total input will be

$$\Delta F_t = \Delta F_1 + \Delta F_2 \tag{8-12}$$

As discussed in Section 2.3 of Chapter 2, three possible observations may be made concerning ΔF_t:

1. $\Delta F_t < 0$.
2. $\Delta F_t > 0$.
3. $\Delta F_t = 0$.

Theoretically, any deviations from the optimum loading would result in an increase in fuel input in dollars per hour. However, for practical purposes, the total cost varies slowly with changes from the minimum cost point, and criterion 3 may be used to a precision within the size of the increment ΔP_1. This criterion becomes precisely correct as ΔP_1 approaches zero. From criterion 3

$$\Delta F_1 + \Delta F_2 = 0 \tag{8-13}$$

$$\Delta F_1 = -\Delta F_2$$

Dividing through by ΔP_1,

$$\frac{\Delta F_1}{\Delta P_1} = -\frac{\Delta F_2}{\Delta P_1} \tag{8-14}$$

From equation 8-11,

$$\Delta P_2 = -\Delta P_1 + \Delta P_{L1,2}$$

Then

$$\frac{\Delta F_1}{\Delta P_1} = -\frac{\Delta F_2}{\Delta P_1}$$

$$= -\frac{\Delta F_2}{\Delta P_2} \frac{\Delta P_2}{\Delta P_1}$$

$$= -\frac{\Delta F_2}{\Delta P_2} \frac{(-\Delta P_1 + \Delta P_{L1,2})}{\Delta P_1} \tag{8-15}$$

Hence

$$\frac{\Delta F_2}{\Delta P_2} = \frac{\Delta F_1}{\Delta P_1} \frac{1}{\left(1 - \dfrac{\Delta P_{L1,2}}{\Delta P_1}\right)} \tag{8-16}$$

As ΔP_1 and ΔP_2 become progressively smaller, we obtain

ALTERNATIVE COORDINATION METHODS

$$\frac{dF_2}{dP_2} = \frac{dF_1}{dP_1} \frac{1}{\left(1 - \frac{dP_{L1,2}}{dP_1}\right)} \tag{8-10}$$

If we choose plant 2 as the reference plant to which all costs are to be referred, then the optimum schedule for a system with j plants is given by solution of the following equations:

$$\frac{dF_2}{dP_2} = \mu = \text{incremental production cost of reference plant 2}$$

$$\frac{dF_1}{dP_1} \frac{1}{\left(1 - \frac{dP_{L1,2}}{dP_1}\right)} = \mu$$

$$\frac{dF_3}{dP_3} \frac{1}{\left(1 - \frac{dP_{L3,2}}{dP_3}\right)} = \mu$$

$$\frac{dF_j}{dP_j} \frac{1}{\left(1 - \frac{dP_{Lj,2}}{dP_j}\right)} = \mu \tag{8-17}$$

where μ = incremental cost at reference plant.

If we denote the reference plant by the subscript n, we obtain

$$\frac{dF_j}{dP_j} \frac{1}{\left(1 - \frac{dP_{Lj,n}}{dP_j}\right)} = \mu = \frac{dF_n}{dP_n} \tag{8-18}$$

or

$$\frac{dF_n}{dP_n} \left(1 - \frac{dP_{Lj,n}}{dP_j}\right) = \frac{dF_j}{dP_j} \tag{8-9}$$

From the above equations it follows that when the system is in economic balance the cost of an increment of power delivered from any of the variable sources to any given point n is the same from each of the variable sources. Of course, the incremental cost of received power at each bus in the system will be different because of the effect of transmission losses.

8.4 SOLUTION OF MODIFIED COORDINATION EQUATIONS THROUGH USE OF MODIFIED INCREMENTAL-LOSS-FORMULA COEFFICIENTS

It is shown in Appendix 8 that $\dfrac{dP_{Lj,n}}{dP_j}$ may be expressed as [3]

$$\frac{dP_{Lj,n}}{dP_j} = 2C_{jk}P_k \tag{8-19}$$

where
$$C_{jk} = K_{jk}'' R_{Gj-Gk} - M_{jk} d_j - N_{jk} f_j \tag{8-20}$$

$$K_{jk}'' = \frac{1}{V_j V_k} [(1 + \Delta s_j s_k) \cos \theta_{jk} + (\Delta s_j - s_k) \sin \theta_{jk}] \tag{8-21}$$

$$M_{jk} = \frac{1}{V_j V_k} (\cos \theta_j + \Delta s_j \sin \theta_j)(\cos \theta_k + s_k \sin \theta_k) \tag{8-22}$$

$$N_{jk} = \frac{1}{V_j V_k} (\sin \theta_j - \Delta s_j \cos \theta_j)(\cos \theta_k + s_k \sin \theta_k) \tag{8-23}$$

The remaining quantities are defined in Chapter 3 and in Appendix 8.

The incremental loss coefficients C_{jk} are more easily calculated than the total loss coefficients B_{mn} and are considered more accurate in the determination of incremental losses between plants.

From equation 8-18 we obtain

$$\frac{dF_j}{dP_j} = \mu \left(1 - \frac{dP_{Lj,n}}{dP_j}\right) \tag{8-24}$$

$$\frac{dF_j}{dP_j} + \mu \frac{dP_{Lj,n}}{dP_j} = \mu \tag{8-25}$$

Equation 8-25 is similar in form to the coordination equations given by equation 5-1 in Chapter 5. The various methods of coordination-equation solution discussed in Chapter 6 may also be applied to equation 8-25.

When equation 8-19 is substituted in equation 8-24 and the straight-line approximation is made for dF_j/dP_j, there results:

$$F_{jj} P_j + f_j = (1 - 2 \sum C_{jk} P_k) \mu \tag{8-26}$$

Solving for P_j, we obtain

$$P_j = \frac{\mu - f_j - 2\mu \sum_{k \neq j} C_{jk} P_k}{F_{jj} + 2\mu C_{jj}}$$

$$= \frac{1 - \dfrac{f_j}{\mu} - 2 \sum_{k \neq j} C_{jk} P_k}{\dfrac{F_{jj}}{\mu} + 2 C_{jj}} \tag{8-27}$$

This equation is identical in form to equation 6-22, and the same programming decks may be used for digital-computer solution as for equation 6–22. Equation 8–27 requires one less P_j calculation than equation 6–22, as P_n is specified when μ is specified. However, total losses are not available as they are when B_{mn} constants are used.

8.5 SOLUTION OF MODIFIED COORDINATION EQUATIONS THROUGH USE OF PHASE-ANGLE RELATIONS [4]

As discussed in Chapter 7, the incremental loss $dP_{Lj,n}/dP_j$ may be simply expressed in terms of voltage phase angles and X/R ratios under certain conditions. For these conditions the modified coordination equations, as given by equation 8–18, may then be solved by an iterative procedure by use of the network analyzer as indicated below:

1. Set up the system on the analyzer as for a load-flow study.
2. For a given total load, set up a trial generation schedule. This first trial schedule would probably be an equal incremental plant production-cost schedule.
3. Read the phase angle of each plant and calculate the difference in phase angle θ_{jn} between each plant and the plant chosen as the reference.
4. Calculate $\dfrac{1}{\left(1 - \dfrac{dP_{Lj,n}}{dP_j}\right)}$ for each plant except the reference plant from a knowledge of the angles given in step 3. If desired, the quantity $\dfrac{1}{\left(1 - \dfrac{dP_{Lj,n}}{dP_j}\right)}$ may be plotted as a function of θ_{jn} and X/R.
5. For economic loading we note from equation 8–18 that

$$\frac{dF_1}{dP_1} \frac{1}{\left(1 - \dfrac{dP_{L1,n}}{dP_1}\right)} = \mu = \frac{dF_n}{dP_n}$$

$$\frac{dF_2}{dP_2} \frac{1}{\left(1 - \dfrac{dP_{L2,n}}{dP_2}\right)} = \mu$$

$$\frac{dF_j}{dP_j} \frac{1}{\left(1 - \dfrac{dP_{Lj,n}}{dP_j}\right)} = \mu$$

With the incremental cost μ at the reference plant known and the value of $\dfrac{1}{\left(1 - \dfrac{dP_{Lj,n}}{dP_j}\right)}$ given in step 4, revised values for all plants except the reference plant may be calculated from the above equations. The operation could be very simply executed on an incremental cost slide rule.

6. With the reference plant as the swing plant, set up this revised schedule on the analyzer and obtain a new set of phase angles.

7. Repeat steps 4, 5, and 6 until the answers have converged.

Several considerations in the application of this method are to be noted:

1. An error is introduced in the general case by application of the approximate formulas of Chapter 7 which relate incremental losses to function of voltage phase angles.

2. Each point of the schedule requires rebalancing of analyzer generators several times to set up successive revisions of schedule. For some systems this step may be very time consuming.

3. Each point of the schedule requires recalibration of all loads.

4. The method is not flexible as it would be necessary to reset the system on the analyzer and restudy if fuel costs or assumed number of units change.

8.6 SOLUTION OF MODIFIED COORDINATION EQUATIONS THROUGH USE OF AUTOMATIC DIGITAL-COMPUTER LOAD FLOWS

The solutions outlined in Sections 8.4 and 8.5 require various approximations to calculate the incremental loss $dP_{Lj,n}/dP_j$. By means of the automatic digital computer load flows may be precisely calculated through the use of iterative methods,[5,6] and it follows that $dP_{Lj,n}/dP_j = \Delta P_{Lj,n}/\Delta P_j$ may also be accurately determined by swinging a small block of generation between points j and n. With the quantities $\Delta P_{Lj,n}/\Delta P_j$ determined, the values of $\dfrac{1}{\left(1 - \dfrac{dP_{Lj,n}}{dP_j}\right)}$ may be calculated for each source. For economic loading it is known that the following condition should be satisfied.

$$\frac{dF_j}{dP_j}\frac{1}{\left(1 - \dfrac{dP_{Lj,n}}{dP_j}\right)} = \mu$$

With the incremental cost μ at the reference bus chosen and with the values of $\dfrac{1}{1 - (dP_{Lj,n}/dP_j)}$ given, revised values for all sources may be

ALTERNATIVE COORDINATION METHODS

obtained. With the reference bus as the swing plant, the system is rescheduled, and the above process is repeated until the answers have converged.

8.7 SUMMARY

An alternative set of coordination equations has been developed. Stating these equations in words, the incremental cost of an increment of power delivered from any of the variable sources to a particular point n should be the same from each source for optimum economy. This may be written mathematically as

$$\frac{dF_j}{dP_j} \frac{1}{\left(1 - \dfrac{dP_{Lj,n}}{dP_j}\right)} = \mu \qquad (8\text{-}18)$$

where

μ = incremental cost at point n in dollars per mw-hr

$\dfrac{dF_j}{dP_j}$ = incremental production cost of source j in dollars per mw-hr

$\dfrac{dP_{Lj,n}}{dP_j}$ = incremental transmission loss in sending an increment of power from source j to point n in megawatts per megawatt

= change in transmission loss divided by change in generation of source j when swinging generation between source j and point n

As will be recalled from Chapter 7, the incremental transmission loss $dP_{Lj,n}/dP_n$ may be approximated by simple expressions involving generator phase angles for transmission systems with elements of similar X/R ratios. For such situations, the use of equations 7–40 and 8–18 to determine generation schedules should be most helpful in studying future system operation and planning on the network analyzer.

APPENDIX 8. DERIVATION OF INCREMENTAL-LOSS-FORMULA COEFFICIENTS [4]

The notation in this appendix is the same as that used in Chapter 3.
If losses are expressed in terms of the reference frame 1 resistances, we obtain

P_L = total transmission losses

$$= i_{Gj}{}^* R_{Gj-Gk} i_{Gk} + i_{Gj}{}^* R_{Gj-Lk} i_{Lk}$$
$$+ i_{Lj}{}^* R_{Lj-Lk} i_{Gk} + i_{Lj}{}^* R_{Lj-Lk} i_{Lk} \qquad (8\text{-}28)$$

Let the subscript n denote the plant used as the reference point in making the reference frame 1 measurements. Then $R_{Gm-Gn} = R_{Gn-Lk} = R_{Lj-Gn} = 0$. Assume that a change Δi_{Gj} at a given plant j is equal and opposite to the change in current that occurs at the reference plant n.[7] Then the change in transmission loss incurred by transferring current from plant j to plant n with all load currents remaining constant is given by:

$$\begin{aligned}\Delta P_{Lj,n} &= \Delta i_{Gj}{}^{*}R_{Gj-Gk}\,\Delta i_{Gk} + \Delta i_{Gj}{}^{*}R_{Gj-Gk}i_{Gk} \\ &\quad + i_{Gj}{}^{*}R_{Gj-Gk}\,\Delta i_{Gk} + \Delta i_{Gj}{}^{*}R_{Gj-Lk}i_{Lk} \\ &\quad + i_{Lj}{}^{*}R_{Lj-Lk}\,\Delta i_{Gk} \\ &= \Delta i_{Gj}{}^{*}R_{Gj-Gk}i_{Gk} + 2\Delta i_{Gj}{}^{*}R_{Gj-Gk}\,\Delta i_{Gk} \\ &\quad + 2\Delta i_{Gj}{}^{*}R_{Gj-Lk}i_{Lk}\end{aligned} \quad (8\text{-}29)$$

Recall that

$$i_{Gj} = i_{dj} + ji_{qj}$$
$$i_{Lk} = i_{Lj}{}' + ji_{Lj}{}''$$

Then equation 8-29 becomes

$$\begin{aligned}\Delta P_{Lj,n} &= \Delta i_{dj}R_{Gj-Gk}\,\Delta i_{dk} + \Delta i_{qj}R_{Gj-Gk}\,\Delta i_{qk} \\ &\quad + 2\Delta i_{dj}R_{Gj-Gk}i_{dk} + 2\Delta i_{qj}R_{Gj-Gk}i_{qk} \\ &\quad + 2\Delta i_{dj}R_{Gj-Lk}i_{Lk}{}' + 2\Delta i_{qj}R_{Gj-Lk}i_{Lk}{}''\end{aligned} \quad (8\text{-}30)$$

Assuming that each load current remains a constant complex fraction of the total load current, as in Chapter 3,

$$d_j = R_{Gj-Lk}l_k{}' \qquad (8\text{-}31)$$
$$f_j = R_{Gj-Lk}l_k{}'' \qquad (8\text{-}32)$$
$$l_k{}' = \Re\,\frac{i_{Lk}}{i_L} \qquad (8\text{-}33)$$
$$l_k{}'' = \mathcal{I}\,\frac{i_{Lk}}{i_L} \qquad (8\text{-}34)$$

Then equation 8-30 becomes

$$\begin{aligned}\Delta P_{Lj,n} &= \Delta i_{dj}R_{Gj-Gk}\,\Delta i_{dk} + \Delta i_{qj}R_{Gj-Gk}\,\Delta i_{qk} \\ &\quad + 2\Delta i_{dj}R_{Gj-Gk}i_{dk} + 2\Delta i_{qj}R_{Gj-Gk}i_{qk} \\ &\quad\qquad\qquad\qquad + 2\Delta i_{dj}d_j i_L + \Delta i_{qj}f_j i_L\end{aligned} \quad (8\text{-}35)$$

Using the relationship $i_L = -\sum\limits_{k} i_{dk}$ with the reference angle chosen

ALTERNATIVE COORDINATION METHODS

so that the summation of source currents equal a real number, we obtain

$$\Delta P_{Lj,n} = \Delta i_{dj} R_{Gj-Gk} \Delta i_{dk} + \Delta i_{qj} R_{Gj-Gk} \Delta i_{qk}$$
$$+ 2\Delta i_{dj} R_{Gj-Gk} i_{dk} + 2\Delta i_{qj} R_{Gj-Gk} i_{qk}$$
$$- 2\Delta i_{dj} d_j i_{dk} - \Delta i_{qj} f_j i_{dk} \quad (8\text{--}36)$$

Define
$$s_j = \frac{Q_j}{P_j} \quad (8\text{--}37)$$

$$\Delta s_j = \frac{\Delta Q_j}{\Delta P_j} \quad (8\text{--}38)$$

The generator currents may be expressed in terms of powers by

$$\Delta i_{dj} = \frac{\Delta P_j}{V_j} (\cos \theta_j + \Delta s_j \sin \theta_j) \quad (8\text{--}39)$$

$$i_{dj} = \frac{P_j}{V_j} (\cos \theta_j + s_j \sin \theta_j) \quad (8\text{--}40)$$

$$\Delta i_{qj} = \frac{\Delta P_j}{V_j} (\sin \theta_j - \Delta s_j \cos \theta_j) \quad (8\text{--}41)$$

$$i_{qj} = \frac{P_j}{V_j} (\sin \theta_j - s_j \cos \theta_j) \quad (8\text{--}42)$$

Equation 8–36 then becomes

$$\Delta P_{Lj,n} = \Delta P_j K_{jk}' + \Delta P_j K_{jk}'' P_k - \Delta P_j M_{jk} d_j P_k - \Delta P_j N_{jk} f_j P_k \quad (8\text{--}43)$$

where

$$K_{jk}' = \frac{1}{V_j V_k} [(1 + \Delta s_j \Delta s_k) \cos \theta_{jk} + (\Delta s_j - \Delta s_k) \sin \theta_{jk}] \quad (8\text{--}44)$$

$$K_{jk}'' = \frac{1}{V_j V_k} [(1 + \Delta s_j s_k) \cos \theta_{jk} + (\Delta s_j - s_k) \sin \theta_{jk}] \quad (8\text{--}45)$$

$$M_{jk} = \frac{1}{V_j V_k} [(\cos \theta_j + \Delta s_j \sin \theta_j)(\cos \theta_k + s_k \sin \theta_k)] \quad (8\text{--}46)$$

$$N_{jk} = \frac{1}{V_j V_k} [(\sin \theta_j - \Delta s_j \cos \theta_j)(\cos \theta_k + s_k \sin \theta_k)] \quad (8\text{--}47)$$

Then,
$$\lim_{\Delta P_j \to 0} \frac{\Delta P_{Lj,n}}{\Delta P_n} = \frac{dP_{Lj,n}}{dP_j} = 2C_{jk} P_k \quad (8\text{--}48)$$

where
$$C_{jk} = K_{jk}'' R_{Gj-Gk} - M_{jk} d_j - N_{jk} f_j \quad (8\text{--}49)$$

References

1. Practical Consideration of Transmission Losses in System Operation, T. W. Schroeder. *AIEE Conference Paper* presented at the 1953 Fall General Meeting, Kansas City, Kansas.
2. Economy Loading Simplified, J. B. Ward. *AIEE Trans.*, Vol. 72, Part III, 1953, pp. 1306–1311.
3. Automatic Digital Computer Applied to Generator Scheduling, A. F. Glimn, R. Habermann, Jr., L. K. Kirchmayer, R. W. Thomas. *AIEE Trans.*, Vol. 73, Part III, 1954, pp. 1267–1275.
4. Coordination of Incremental Fuel Costs and Incremental Transmission Losses by Functions of Voltage Phase Angles, W. R. Brownlee. *AIEE Trans.*, Vol. 73, Part III, 1954, pp. 529–533.
5. Automatic Calculation of Load Flows, A. F. Glimn, G. W. Stagg. *AIEE Trans. Paper* presented at the Summer General Meeting, June 24–28, 1957, Montreal, Quebec, Canada.
6. Extensions in Digital Load Flow Techniques Using High-Speed Computer, M. S. Dyrkacz, A. F. Glimn, D. G. Lewis. *AIEE Conference Paper* presented at the Fall General Meeting, October 9, 1957, Chicago, Illinois.
7. Power Losses in Interconnected Transmission Networks, H. W. Hale. *AIEE Trans.*, Vol. 71, Part III, 1952, pp. 993–998.

Problems

Problem 8.1

Assume that the plant incremental-production-cost data may be approximated by

$$\frac{dF_1}{dP_1} = 2 + (1.0)P_1 = F_{11}P_1 + f_1$$

$$\frac{dF_2}{dP_2} = 1.5 + (1.0)P_2 = F_{22}P_2 + f_2$$

where P is expressed in per unit on a 100-mva base. Assume that the maximum load of each unit is 100 mw; the minimum load, 10 mw.

Assume that the losses incurred in swinging generation between plant 1 and plant 2 are as tabulated:

Total Transmission Losses in p.u.	P_1 in p.u.
0.1008	0.2
0.0785	0.3
0.0605	0.4
0.0470	0.5
0.0378	0.6
0.0331	0.7
0.0327	0.8
0.0368	0.9
0.0452	1.0

Find the most economic allocation of generation for a total received load of 1.20 p.u.

Problem 8.2

Assume that the system under consideration is in economic dispatch and that the incremental cost of source 1 is three dollars per mw-hr and the incremental cost at source 2 is four dollars per mw-hr. For this condition, what would be the incremental loss in transferring power from source 1 to source 2?

9 EVALUATION OF ENERGY DIFFERENCES IN THE ECONOMIC COMPARISON OF ALTERNATIVE FACILITIES

9.1 INTRODUCTION

This chapter is limited to a brief discussion of the evaluation of energy differences in the economic comparison of alternative facilities [1,2] and does not discuss the evaluation of investment differences. This discussion is of particular value in the study of alternative plant locations and alternative transmission facilities.

9.2 METHOD OF ANALYSIS

In a widespread system the scheduling of generation of a given plant at a given site is dependent upon its thermal characteristics, the cost of the fuel at the site, and the electrical location in the network. In order to compare the system operation for alternative plans it is necessary to determine the optimum operation of the system for each alternative plan. The final step in the analysis would then involve determining the difference in operating costs for the two alternatives.

A brief summary of the steps in such an analysis follows:

1. Determine transmission-loss formulas for the transmission systems for the alternative plans.

2. By methods described in Chapters 5 and 6 for coordinating incremental production costs and incremental transmission losses determine generation schedules for various total loads for both alternatives. These total loads are determined by inspection of the load-duration curve.

3. Determine the transmission losses for the generation schedules by application of the transmission-loss formulas.

4. Calculate the total fuel input for all the plants in the system in dollars per mw-hr received as a function of received load for both alternatives.

5. Subtract the two fuel inputs in dollars per mw-hr received in step 4 and obtain a curve of the savings in dollars per mw-hr received as a function of received load.

6. The results of step 5 may then be combined with the load-duration curve as described in Chapter 6 to obtain the annual savings involved.

It is particularly helpful in conducting such analysis to make use of a digital computer, as described in Section 4.8. The digital computer, in addition to calculating the allocation of generation for optimum economy, which could also be obtained from an analogue-type dispatching computer, may also be programmed to calculate automatically such valuable information as total losses, received load, fuel input to each unit, and total fuel input. By extending the digital-computer flow diagram of Figure 6.6 in a manner suggested in Section 6.5, the fuel inputs for alternative plans may be obtained for the same desired values of received loads. The resulting digital-computer flow diagram is given in Figure 9.1. The desired received load point is designated as P_R^d. The other symbols may be reviewed by reference to Chapter 6. In step 9 of Figure 9.1 the tolerance chosen in matching the actual received load to the desired load is usually taken as somewhere between 0.1 and 0.01 mw, depending on the size of the system and the accuracy required. By calculating fuel inputs for the same received loads, the fuel-cost differential may be obtained by direct subtraction of the fuel inputs for various values of received load for the alternative plans.

The steps involved in calculating loss formulas for the alternatives may be omitted if the power load flows are obtained digitally by such methods as described in References 3 and 4. With the load flow thus established, schedules may be obtained without the use of a loss formula by the method given in Section 8.6. The total transmission losses are given directly by subtracting the sum of the loads from the sum of the generation. As a check, the computer may also be programmed to calculate the losses by summing the I^2R losses, line-by-line. Another method of obtaining an economic schedule, using digital computers, is given in Reference 5. The two all-digital methods suggested here do not require the approximations involved in calculating a loss formula.

9.3 EXAMPLE OF METHOD

This particular example considers the problem of the location of the next 95-mw unit to be installed on a given power system. Two locations are available. One location is at plant A; the other, at plant O. Generation schedules corresponding to the two unit locations are given in

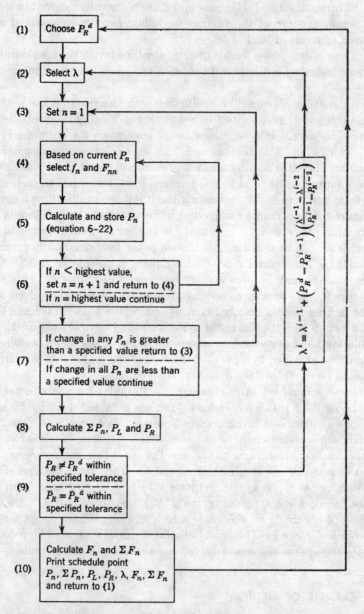

Figure 9.1. Digital computer flow diagram for calculation of optimum schedule and fuel input for given received load.

Figures 9.2 and 9.3. The corresponding transmission losses are given in Figure 9.4. By application of steps 4, 5, and 6 of Section 9.2, the following savings are obtained:

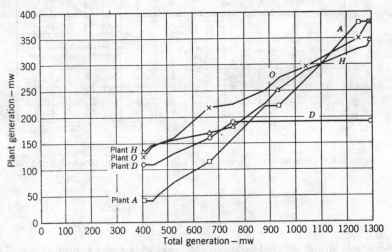

Figure 9.2. Generation schedule with next 95-mw unit at plant A.

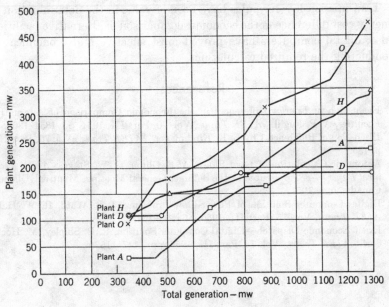

Figure 9.3. Generation schedule with next 95-mw unit at plant B.

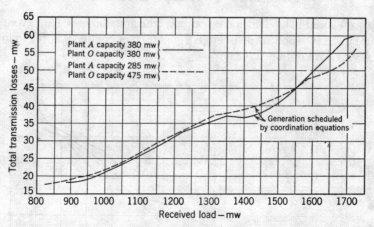

Figure 9.4. Comparison of transmission losses for alternative unit locations.

Annual fuel savings obtained by locating the next 95-mw unit at plant O instead of plant A.

9.4 SUMMARY

Transmission-loss formulas have been of value in the determination of energy cost difference in the economic comparison of alternative facilities. Also, digital computers have proved most effective and economical in conducting the required calculations.

References

1. Evaluation of Energy Differences in the Economic Comparison of Alternative Facilities, A. P. Fugill. *AIEE Trans.*, Vol. 70, Part III, 1951, pp. 1854–1860.
2. Discussion of Reference 1 by L. K. Kirchmayer. *AIEE Trans.*, Vol. 70, Part III, p. 1859.
3. Automatic Calculation of Load Flows, A. F. Glimn, G. W. Stagg. *AIEE Technical Paper 57-681* presented at the Summer General Meeting, Montreal, Quebec, Canada, June 1957.
4. Digital Computer Solution of Power-Flow Problems, J. B. Ward, H. W. Hale. *AIEE Trans.*, Vol. 75, Part III, 1956, pp. 398–404.
5. Exact Economic Dispatch—Digital Computer Solution, R. B. Shipley, M. Hochdorf. *AIEE Trans.*, Vol. 74, Part III, 1956, pp. 1147–1152.

INDEX

Accuracy requirements, 28
Alternative coordination methods, 238
American Gas and Electric System, 3, 4, 38, 135, 137, 165, 170, 179, 209
Analogue computers, *see* Differential analyzer, Dispatching computer, Network analyzer, Penalty-factor computer, Simultaneous equation solver
Analogue solution of coordination equations, 188
Angle, transmission losses as function of, 218
Approximation of incremental costs, 14
Arrow rule of multiplication, 59
Automatic control of interconnected areas, 4
Automatic digital computer, *see* Digital computer
Automatic dispatching system, 210
  savings from, 215
Auto-transformer, 98

Balancing flows, 152
Base case load flow, 118, 119
Branch flows, 154
Branch impedances, 152

Calculation of, A_{mn}, 125
  B_{mn}, 127
  B_{no}, 96, 97, 148
  B_{oo}, 96, 97, 148
  d_n, 120
  f_n, 123
  H_{mn}, 125
  K_{mn}, 124
  w', 127

Capacity commitments, 27, 204
Card-programmed calculator, 138
Card punch, 144
Card reader, 144
Circuit, new, 62
  old, 62
Circuit transformation, 62
Computer, *see* Differential analyzer, Digital computer, Dispatching computer, Electronic data-processing machine, Network analyzer, Penalty-factor computer, Simultaneous equation solver
Computer flow diagram, *see* Digital computer flow diagram
Constant loads, 96
Coordination equation application to two-plant system, 165
Coordination equations, approximate linear, 162
  approximate penalty-factor, 163
  derivation of, 182
  digital computer solution of, 193
  electronic differential analyzer analogue of, 191
  exact nonlinear, 161
  load-dispatching computer solution of, 211
  mesh-circuit analogue of, 188
  nodal-circuit analogue of, 189
  physical interpretation of, 163
Coordination methods, 161
  alternative, 238
Current, turn-ratio, 102

Data address, 141
Determinant, 193

INDEX

Differential analyzer, 189, 192
Differential analyzer coordination equation analogue, 191
Digital computer, 2, 4, 117, 141, 148, 157, 199, 215, 246, 256
  results from, 142
  schematic representation of, 138
Digital computer flow diagram, 139, 199
  coordination equation, 200
  loss formula, 140, 149
  optimum schedule and fuel input for given received load, 254
  self and mutual impedance calculation, 156
Digital computer load flows, 120, 246, 253
Digital computer programming, 139, 141
Digital computer storage, 138, 141, 144
Digital voltmeter, 213
Dispatching computer, 205, 210
Dispatching procedure, manual, 28
Driving point and transfer impedances, 218

Economic automation, 4, 187
  savings from, 215
Economic dispatching computers, 210
Electrical system data, 186
Electronic data-processing machine, 143, 144
Electronic differential analyzer, *see* Differential analyzer
Element of matrix, 58
Equation of constraint, 182
Equivalent load current, 67, 119
Equivalent transmission loss circuit, 48
Errors in economic dispatching, 28, 37
Evaluation, of energy differences, 252
  of savings, 187, 202
Exact coordination equations, 161

Flow diagram, *see* Digital computer flow diagram
Fuel input, 8
Function generator, 212, 213

General loss formula, 144
  additional savings obtained, 179

Heat rate, 8, 9

Hydroelectric power system, 3
Hypothetical load current, 73, 102
Hypothetical load point, 74

Identification indices, 52
Imaginary components of load currents, 128
Impedance, self and mutual, 52, 53
  driving-point and transfer, 218
Impedance transformation, 63
Incremental cost, of delivered power computer, 212
  of labor, 10
  of maintenance, 10
  of received power, 10, 161, 163
  of supplies, 10
Incremental-cost slide rule, 3, 13, 14, 205, 206, 246
Incremental efficiency, 164, 182
Incremental fuel cost, definition, 9
Incremental fuel cost slide rule, *see* Incremental-cost slide rule
Incremental fuel rate, definition, 8
Incremental fuel rate slide rule, *see* Incremental-cost slide rule
Incremental loss, to hypothetical system load, 161, 163
  in swinging generation between two points, 220
Incremental production costs, definition of, 10
  error in representation, 28, 37
  straight-line approximation of, 14
Incremental transmission loss formula, 161, 180, 222, 226, 233, 235, 243
Index, repeated, 50
Index notation, 50, 59, 60
Indices, identification, 52
Input-output curve, 8
Interconnected areas, control of, 4
Interconnection of companies, 1
Interconnection systems accounting, 4
Interconnection transactions, 1
Intermediate generation, 223
Intermediate loads, 223
Internally programmed digital computers, *see* Digital computer
Invariant transformations, 62
Inverse of a matrix, 62, 193

INDEX

Iterative digital method, advantages of, 201
Iterative methods, 4, 196, 198, 245, 246

Labor, incremental cost of, 10
Lagrangian multipliers, 40, 182
Least squares, method of, 144
Load-dispatching computers, 210
Load flows, 118–120, 246, 253
Load pick-up ability, 27
Loads, constant, 96
   intermediate, 223
   nonconforming, 95, 110
Load variation, 180
Location of generating units, 253
Loop-admittance matrix, 153
Loop flows, 154
Loop-impedance matrix, 153
Loop-voltage drops, 152
Loss of economy, calculation of, 41
   resulting from error in holding generation at desired value, 37
   resulting from error in incremental cost representation, 37
Losses, reactive, 97
   real, 81

Magnetic drum, 141
Magnetic tape unit, 144
Main diagonal, 83
Maintenance, incremental cost of, 10
Matrix, conjugate of, 63
   current, 50
   element of, 58
   impedance, 50
   inverse, 93
   skew symmetric, 83
   symmetric, 83
   transformation, 63
   transpose of, 60
   unit, 61
   voltage, 50
Matrix inversion, 193
Matrix multiplication, 58
Matrix notation, 49
Mesh-circuit analogue, 188
Minor, 194
Modified coordination equation, 239
   solution by digital load flows, 244, 246

Modified incremental loss formula, derivation of, 247
   form of, 243
Mutual impedances between generators and loads, 53, 54
Mutual transformers, 189

Network analyzer, 2, 3, 48, 116, 117, 188, 238
Network-analyzer calculation of schedules, 3, 238
Network-analyzer mesh-circuit analogue of coordination equations, 189
Network-analyzer nodal-circuit analogue of coordination equations, 189
Niagara Mohawk System, 232
Nonconforming loads, 95, 110
Normal load data for loss formula, 119

Off-nominal transformer ratios, 98, 153
Open-circuit impedance measurements, 116
Operation instruction, 141
Optimum economy, 181, 247
Optimum scheduling, derivation of equations for, 182, 240

Penalty factor, 3, 163, 182
   simple example of, 164
Penalty-factor computer, 4, 205
   application of, 208, 210
   circuitry for, 209
Phase-angle transmission loss formulas, 218, 226, 233
   applicability, 226, 231
   comparison with B_{mn} formula, 231
   derivation for general case, 235
   effect of intermediate generation, 228
   effect of intermediate load, 226
   effect of X/R ratios, 231, 235
   use for solving modified coordination equations, 245
Pivotal condensation, 195
Plant data, 186
Power-angle equations, 218
Precalculated generation schedules, 187, 205
Printer, 144
Punched cards, 3, 138, 141
Push-button selectors, 210

Quadratic form, 59

Reactive characteristics, 94, 121, 131, 181
Reactive losses, 97
Reference frame, definition of, 52
Reference frame 1, 52
Reference frame 2, 67
Reference frame 3, 74
Reference frame 4, 89
Reference frame 5, 89
Reference frame 6, 89
Repeated index, 50
Representation, of loads, 95, 110
   of reactive characteristics, 94, 121, 131, 181
Reserve requirements, 27
Residual, 144
Reversing switch, 209

Savings, by economic automation, 4, 215
   procedure for evaluation of, 202
   by transmission loss considerations, 1, 177
Schedules, analogue calculation of, 188
   digital calculation of, 193
   by incremental-cost slide rule, 13
   by penalty-factor computer, 205
   precalculated, 187, 205
   by special dispatching computers, 205, 210
   by successive tests on analyzer, 238
Scheduling of turbine generators, by base loading to capacity, 27
   by base loading to most efficient load, 27
   by equal incremental cost of received power, 246
   by equal incremental costs, 10
   by equal incremental costs at point n, 243, 247
   by loading proportional to capacity, 27
Self and mutual impedances, 52
   computer time, for determination of, 139, 157
   between generators, 53
   between generators and loads, 53

Self and mutual impedances, between loads, 53
Simultaneous equation solver, 193
Skew-symmetric matrix, 83
Skew-symmetric reference frame 3 reactance, 82, 88
Slide rule, see Incremental-cost slide rule
Stability limitations, 27
Starring, 195
Supplies, incremental cost of, 10
Symmetric matrix, 83
Symmetric reference frame 3 resistance, 82

Tap settings, 98
Total turn-ratio current, 102
Tracks, 148, 150
Transformation, of circuits, 63
   concept of, 62
   impedance, 63
   laws of, 63, 110
   voltage, 63
Transformation matrix, 62
Transformer, variable, 209
Transformer ratio, 151
Transmission-loss formula, assumptions, 110
   calculation of, 116
   development, 48
   form of, 2, 3
   historical review of development, 2, 3
Tramission-loss phase-angle formulas, see Phase-angle transmission-loss formulas
Transmission losses, effect on fuel economy, 37, 177
Trigonometric projections, 90
Turn ratios, 98

Unit commitments, 27, 204
Unit matrix, 61

Valve discontinuities, 14
Variable transformer, 209
Voltage limitations, 27
Voltage transformation, 63

X/R ratios, 2, 218, 231, 235